Cooperating with Nature

Confronting Natural Hazards with Land-Use Planning for Sustainable Communities

Raymond J. Burby, *Editor*

JOSEPH HENRY PRESS
Washington, D.C. 1998

JOSEPH HENRY PRESS • 2101 Constitution Avenue, N.W. • Washington, D.C. 20418

The Joseph Henry Press, an imprint of the National Academy Press, was created with the goal of making books on science, technology, and health more widely available to professionals and the public. Joseph Henry was one of the founders of the National Academy of Sciences and a leader of early American science.

Library of Congress Cataloging-in-Publication Data

Cooperating with nature : confronting natural hazards with land use planning for sustainable communities / editor, Raymond J. Burby : authors, Timothy Beatley . . . [et al.].
p. cm. — (Natural hazards and disasters)
Includes bibliographical references and index.
ISBN 0-309-06362-0 (alk. paper)
1. Natural disasters—United States—Evaluation. 2. Land use—United States—Management. 3. Community development—United States. 4. Sustainable development—United States. I. Burby, Raymond J., 1942- . II. Beatley, Timothy, 1957- . III. Series.
GB5010.C67 1998
363.346—dc21 98-12415
CIP

Printed in the United States of America.

Foreword

THE NATION'S FIRST NATURAL hazards assessment got under way in 1972 at the Institute of Behavioral Science at the University of Colorado. Funded by the National Science Foundation and led by geographer Gilbert White and sociologist J. Eugene Haas, it was an interdisciplinary effort involving scores of policymakers, practitioners, and scholars from across the nation. Its purpose was to assess our knowledge about natural hazards and disasters, to identify major needed policy directions, and to inventory future research needs. The volume summarizing this endeavor, *Assessment of Research on Natural Hazards in the United States,* published in 1975, was a landmark in what was then a new field of study.

A quarter of a century later we find ourselves in a national conversation about how natural hazards mitigation can result in disaster-resilient communities. This conversation began in the early 1990s among a few individuals working in federal agencies and academia, and was articulated at a workshop in Estes Park, Colorado, in 1992, which was attended by many of the nation's leading natural hazards experts. They concluded that a second assessment of hazards in the

United States was needed, and that its unifying theme should be sustainable development, or development that enhances the capacity of the planet to provide a high quality of life now and in the future. A subsequent workshop in 1994 brought many of the same experts together to discuss and formulate an agenda for this second hazards assessment.

This book, which ensues from the second natural hazards assessment, is one of a series of works on natural hazards and disasters published by the Joseph Henry Press. A select group of experts were invited to expand upon their necessarily condensed contributions to the second assessment by developing individual works on major themes in the natural hazards and disasters field, including insurance, risk assessment, disaster preparedness and response, and mapping. This volume on land use management and natural hazards was written by a group headed by Raymond Burby of the College of Urban and Public Affairs at the University of New Orleans. *Cooperating with Nature* focuses on the breakdown in sustainability that is signaled by natural disaster. The authors chronicle the long evolution of land use planning and identify key components of sustainable planning for hazards. The book describes the promise of land use management for achieving sustainability, explores reasons why this promise is not being realized uniformly by government at various levels, and proposes ways to foster sound land use decision making. The authors explain why sustainability and land use have not been taken into account in the formulation of public policy. They articulate a vision of sustainability, giving concrete suggestions for policy reform, and calling for a new National Hazardous Area Management Act and program to foster improved planning and management at state and local levels.

The second national natural hazards assessment was funded by the National Science Foundation under Grant Number 93-12647, with supporting contributions from the Federal Emergency Management Agency, the U.S. Environmental Protection Agency, the U.S. Forest Service, and the U.S. Geological Survey. The support of these agencies is greatly appreciated. However, only the authors are responsible for the information, analyses, and recommendations contained in this book. Very special thanks are extended to J. Eleonora Sabadell and William A. Anderson of the National Science Foundation for placing their confidence in us to carry out this mission.

DENNIS S. MILETI, *Director*
Natural Hazards Research and
Applications Information Center
University of Colorado at Boulder

Contributors

RAYMOND J. BURBY, *Editor,* is DeBlois Chair of Urban and Public Affairs at the University of New Orleans. Former co-editor of the *Journal of the American Planning Association*, he is the author and editor of 14 books and numerous articles dealing with environmental management and planning. His research focuses on environmental hazards, regulatory systems, and local planning and land-use management.

Contributing Authors

Timothy Beatley, University of Virginia
Philip R. Berke, University of North Carolina at Chapel Hill
Raymond J. Burby, University of New Orleans
Robert E. Deyle, Florida State University
Steven P. French, Georgia Institute of Technology
David R. Godschalk, University of North Carolina at Chapel Hill
Edward J. Kaiser, University of North Carolina at Chapel Hill
Jack D. Kartez, University of Southern Maine
Peter J. May, University of Washington
Robert B. Olshansky, University of Illinois
Robert G. Paterson, University of Texas
Rutherford H. Platt, University of Massachusetts

Contents

Cooperating with Nature

CHAPTER ONE

Natural Hazards and Land Use: An Introduction

RAYMOND J. BURBY

THIS BOOK IS ABOUT NATURAL disasters and sustainability—the capacity of the planet to provide a high quality of life for not only present but also future generations. Hardly a month passes without screaming headlines informing readers of the latest natural catastrophe that has befallen some part of the world. Newspaper subscribers might also notice indications, buried in the Op/Ed section, that scientists are becoming concerned about our planet's capacity to sustain a high quality of life. Disasters signal a serious breakdown in sustainability. The earthquakes, floods, hurricanes, and other natural events that cause so much concern have always been present. But now disasters wreak havoc that goes well beyond the ability of society to take them in stride. What is more, in trying to reduce vulnerability by taming nature, the situation is often made worse, not better. But there is a better way.

By planning for and managing land use to enhance sustainability, we can reduce our vulnerability to disasters, if not eliminate them. Land use plans enable local governments to gather and analyze information about the suitability of land for development, so that the limitations of hazard-prone areas are understood by policymakers, potential investors, and community residents. In the process of preparing plans, local governments engage in a problem-solving process that works to ensure that all stakeholders understand the choices the community faces, and that they reach some degree of consensus about how these choices will be made. For example, should existing areas at risk of flooding be protected by levees or evacuated and maintained as open space? Should community growth be channeled away from hazardous areas or do economic benefits from their development justify the expense of protection? How can protection be provided most effectively? Plans require the systematic evaluation of alternative courses of action, so that the approach chosen to reduce vulnerability is an optimal solution given a community's present circumstances, future prospects, and the goals and aspirations of its residents.

While plans provide general guidance for managing development, land use regulations such as zoning set specific rules for the private sector about where development will be allowed and how development (and redevelopment) is to take place so that vulnerability from natural hazards is minimized. Examples of safe development practices include elevating buildings and transportation infrastructure above expected flood heights, and strengthening buildings to withstand ground shaking in an earthquake or high winds in a hurricane. Land use management also includes complementary governmental policies, such as campaigns to inform citizens about areas that face the greatest risks from hazards, and policies to locate infrastructure such as roads and sewer lines so that they steer development away from hazardous areas. In combination, plans and land use management programs enhance prospects for a sustainable future—one in which citizens and their elected officials make informed choices about using hazardous areas in ways that will not jeopardize the long-term viability of their communities.

The logic of land use planning and management, on its face, seems self-evident. But, for a variety of reasons few local governments, of their own accord, pay much attention to hazards before disaster strikes. The federal government and a few states realize the importance of proactive programs to reduce risk, but they have been reluctant to force local governments to employ land use planning and management techniques. As a result, unsustainable development practices (and large losses in natural disasters) have been the rule rather than the exception in urban areas across America.

In this book we highlight the promise of land use planning and management for achieving sustainability, explore reasons why this promise is not being realized uniformly by local governments, and propose ways to foster sound land use decision making. We will discuss why land use and sustainability have been ignored in devising policy to deal with natural hazards, and lay out planning procedures, a vision of sustainability, and concrete suggestions for policy reform.

WHY NATURAL DISASTERS NEED ATTENTION

Natural disasters exact a heavy toll, and it is increasing every year. Worldwide more than three million people have lost their lives in flood, earthquake, landslide, hurricane, volcanic, and other natural disasters over the past two decades, and almost one billion have been affected in other ways. The economic consequences have been devastating. Property damages have been estimated to range from $25 to $100 billion between 1965 and 1985 (Advisory Committee on the International Decade for Natural Hazard Reduction, 1987). But losses are obviously much higher than that since disasters cause extensive disruption to economic systems. In fact, one source put the total economic costs of disaster in the early 1990s at more than $100 billion per year (Clarke and Munasinghe, 1995). In any given year, however, they can be much greater. The Kobe earthquake of January 17, 1995, produced this toll of human suffering: 6,308 deaths, 43,177 injuries, 300,000 people homeless, and $100 billion in total property damages.

In the United States property losses also have been extraordinary in the 1990s: for example, $30 billion in Hurricane Andrew in 1992; $16 billion in flooding in the upper Mississippi and Missouri River basins in 1993; $20 billion in the Northridge earthquake in 1994; $5.5 billion in Hurricanes Luis and Opal in 1995; and more than $1 billion in Hurricane Fran in 1996. Most of these losses were absorbed by households and businesses, but the public costs are substantial, too. Congress appropriated more than $115 billion for disaster relief and mitigation between 1975 and 1995. State and local governments absorbed billions more in losses. Although accurate loss data for developing nations are difficult to secure, one estimate indicates that, as a percent of gross national product, losses are almost 20 times higher in developing countries than in developed countries (Burton et al., 1993).

Grim as these statistics are, it is possible that society so far has seen only the tip of an enormous iceberg of risk. Forecasts of potential future losses lurking beneath the surface of the sea of public complacency regarding natural hazards are truly frightening. Consider these worst-case earthquake scenarios for Los Angeles, San Francisco, and Tokyo developed by Shah (1994). A magnitude 7.0 earthquake on the Newport-Inglewood Fault in Los Angeles could cause $170 to $220 billion in economic losses, $95 to $120 billion in insured losses, 3,000 to 8,000 deaths, and up to 20,000 serious injuries. A repeat of the 1906 earthquake in San Francisco (a magnitude 8.3 event) could result in $170 to $225 billion in economic losses, $80 to $105 billion in insured losses, 3,000 to 8,000 deaths, and up to 18,000 serious injuries. A repeat of the 1923 Tokyo earthquake would cause economic losses of $2.0 to $3.3 trillion, insured losses of $30 to $40 billion, 40,000 to 60,000 deaths, and serious injuries in the hundreds of thousands.

WHY PAST WAYS OF COPING ARE INADEQUATE

Governments traditionally try to cope with disasters using warnings before disaster strikes, emergency relief after a disaster occurs, and hazard reduction measures such as levees to reduce the likelihood of a future disaster. For a variety of reasons, none of these usual approaches is adequate to reduce losses in disasters to acceptable levels. In addition, each is enormously expensive.

Too Few Heed Warnings

Warning is the oldest way of coping with disaster, and it is still ubiquitous. Some warning systems try to reduce losses immediately preceding the onset of a hazardous event. Home phones that ring automatically to sound the alarm when flood waters begin to rise rapidly are an example. Others try to reduce losses well before a disaster occurs by letting people know that particular areas are affected by hazards. Examples include maps that delineate flood and seismic hazard zones, signs posted to warn of flood hazards, and laws that require real estate agents to inform their clients of hazards. More informative warnings are based on hazard assessments that not only identify the presence of natural hazards but also provide information on the amount of loss that may be incurred on average every year, or for a specified scenario such as a 100-year flood (a flood that has a 1 percent chance of occurring in any given year).

Warning is an essential ingredient in any strategy to cope with losses from disasters, since no method of coping—even the land use strategies we argue for in this book—can completely eliminate risk. At the very minimum, therefore, people need to be informed of the risks they face so they can make well-considered decisions based on their own calculations of costs and benefits. But warning has weaknesses, so it cannot be relied upon as the sole means of coping.

A variety of psychological factors lead people to ignore immediate warnings of imminent disaster as well as cautions about using land that is hazardous. Some people are not able to accurately perceive probabilities of loss, even when they have been told a hazard exists. As a result, they tend to heavily discount any benefits from avoiding a hazard or taking action to reduce vulnerability. Even after they have been proven wrong and a disaster occurs, people tend to assume it will never recur (a conclusion likely to be refuted by subsequent events). Dissonance theory (i.e., beliefs follow behavior) predicts that, once people have exposed themselves to a hazard by locating in an area at risk, they will assume any hazards that might exist are trivial. Also, some people become fatalistic when thinking about hazards when they can neither prevent them nor estimate (very accurately) when and where losses are likely to occur. Thus, psychological difficulties in perceiving hazards lead most people to simply ignore or greatly discount the potential for disaster.

Difficulty in accurately estimating risk also limits the effectiveness of warnings. Hurricane Andrew offers an excellent example. In 1989 the South Florida Regional Planning Council published a vulnerability study that estimated maximum probable losses in south Florida at about $5 billion from a worst-case, Category 5 hurricane. Losses actually incurred three years later from Andrew, which was a lower intensity Category 4 hurricane, were five to six times larger! Geographers Ian Burton, Robert Kates, and Gilbert White (1993), in fact, have observed that experts frequently err on the low side when estimating vulnerability to losses from natural hazards. As a result, reliance on warnings may lead people to expose themselves and their property to a greater degree than they would have, had information about hazards been more accurate.

Relief and Insurance Can Also Foster Excessive Exposure to Risk

The limitations of warnings early on led governments to look to other means of coping with disasters. Relief and insurance, which take effect after catastrophic losses occur, reduce the adverse impacts of disaster and ease reconstruction and recovery. Research undertaken in the late 1970s provided strong evidence that relief and other emergency measures work; studying communities some years after they had been hit by disasters, researchers could find no evidence of adverse effects (see Friesma et al., 1979; Wright et al., 1979).

But relief, too, has its problems. Lack of adequate preparation, planning, and coordination of services are endemic, so that in some cases the relief effort (particularly by subnational governments) is inadequate, and short-term suffering can be acute, as was the case in south Florida after Hurricane Andrew. Because it subsidizes people and firms occupying hazardous areas, relief can produce complacency. If they believe someone else will pick up the tab, individuals and communities may not be willing to take the steps necessary to reduce their own vulnerability, even when such steps are feasible and proven to be cost effective. This complaint is frequently voiced by economists (see, for example, Lichtenberg, 1994). And, of course, the costs are astronomical, as the billions of dollars now appropriated every year for disaster relief in the United States demonstrate. Furthermore, relief does little to prevent recurrence of disasters, since people frequently rebuild in the same disaster-prone locations using the same unsafe building techniques.

Insurance compensates for losses in a way that provides greater assurance to property owners and avoids some of the pitfalls of relief. If insurance rates are set at levels that actually reflect the risk of loss, then insurance premiums can transmit accurate economic signals to policyholders. As risk (and premium rates) go up, some people and firms will decide that the benefits of locating in (or continuing to occupy) a hazardous area are not worth the added insurance costs they will have to bear, and therefore they will locate elsewhere. Even those who have already located in a hazardous area and cannot relocate may adopt loss reduction measures such as elevating structures in a floodplain or strengthening buildings susceptible to shaking in an earthquake, if in doing so they can reduce their insurance rates. However, insurance also has a number of limitations.

Because of the potential for catastrophic losses, insurance providers must have the ability to spread losses widely over a broad area or to draw on reinsurance markets or public financial assistance. Spreading losses over a wide area has proven virtually impossible, however, because the psychological barriers to risk perception noted earlier lead most people to view insurance against natural disasters as a poor value. Only those most at risk—and whose frequent losses would drive a company insuring their property into bankruptcy—willingly purchase such insurance. In the nine-state region affected by the 1993 floods in the Midwest, for example, only about 20 percent of the structures in the floodplain were insured, a rate that is typical of the United States as a whole (see Interagency Floodplain Management Review Committee, 1994). Similar findings have been reported for earthquake insurance in California, where only about one in five households is insured.

The enormous insurance losses in recent natural disasters make reinsurance very difficult for companies to find. In response to heavy outlays required by recent disasters, the insurance industry has tried to limit catastrophic losses of reserves by shifting risk to government. When insurance companies began canceling policies in Florida following Hurricane Andrew, for example, the State of Florida instituted a surcharge on property insurance policies to create the Hurricane Catastrophe Fund, a reinsurance fund for private insurers. Florida and Texas have created "wind pools" to provide insurance coverage to property owners who otherwise would not be able to afford coverage. Finally, an effort has been made to shift more hazard insurance risk to the federal government. Bills introduced in Congress in 1994 and 1995 would have established a public corporation to provide all-hazards insurance, with the federal treasury providing reinsurance for catastrophic losses that exceed loss reserves. Congress, however, has not passed legislation that would have the federal government expand its the role from that of an insurer of risk from flooding to a re-insurer of risks from all natural hazards.

Structural Hazard Reduction Can Be Costly

Areas susceptible to natural hazards often have attributes that make them attractive for economic use. Floodplains and volcanic soils, which are particularly fertile, are farmed intensively. Most major cities were founded and initially grew because of their proximity to rivers that provided a ready means to import raw materials and export goods to far-flung markets. Those same floodplains in more recent times have been favored routes for expressways, because little effort and expense is needed for clearing and grading rights-of-way. The economic advantages provided by river and highway access have attracted billions of dollars of investment in areas at risk of flooding. In coastal areas at risk from hurricanes and winter storms, a variety of economic advantages have attracted development. Coastal fisheries, transportation through ports and along intracoastal waterways, and outdoor recreation have each attracted millions of people and billions of dollars in investment. These large investments in areas at risk from natural hazards early on led engineers and policymakers to look for ways to reduce risk while continuing to reap the rewards of vulnerable locations.

Throughout history a variety of engineering methods have been used to keep hazards such as riverine and coastal flooding away from people and damageable property. Dams and flood storage reservoirs, levees, dikes, pumps, channel improvements and diversions, sea walls, and groins—each of these measures provides protection for large areas, and thus can be both effective and efficient in reducing losses.

The benefits to be gained from controlling the hazard, however, have to be weighed against the serious costs and various shortcomings of the structural approach. Structural protection against natural disasters comes at a high cost in dollars and does not provide complete protection, as massive flood damages to tens of thousands of homes and businesses behind breached levees in the upper Mississippi and Missouri River basins tragically demonstrated in 1993. The failure of structural protection in the 1993 flooding is not an isolated occurrence. A report prepared for the U.S. government indicates that fully two-thirds of national losses in flooding result from catastrophic events that exceed the design limitations of engineering works that are relied on to provide safety (Sheaffer et al., 1976).

In part, disasters such as the Midwest flooding in 1993 stem from a curious side effect of structural protection. People and businesses tend to view the structures as providing *complete* protection from loss, when in fact they provide only *partial* protection up to the limits of some storm event, such as a storm that happens, on average, once in 50 or 100 years. Even lower magnitude storms may overwhelm structural protection as a result of incorrect design, operational mistakes, or loss of capacity, such as the silting up of flood storage reservoirs. Because people do not understand that structural protection has limits, however, structures have been found to actually induce development in hazardous areas and to increase, not decrease, the likelihood that when a large flood or hurricane does occur, losses truly will be catastrophic (for evidence of this effect, see White et al., 1958, and Burby and French et al., 1985). Local governments have contributed to this loss scenario by waiving requirements for the elevation of buildings located behind levees, apparently because, like individuals, they overestimate the degree of protection structures actually provide.

In addition to catastrophic losses of life and property behind engineered structures, incalculable additional losses to nature have sometimes accompanied structural efforts to reduce vulnerability. For example, levees along hundreds of miles of the Mississippi River erected by the U.S. Army Corps of Engineers following flooding in the late 1920s have, in turn, contributed to accelerated erosion in the state of Louisiana because the river's silt no longer replenishes wetlands. The conversion of wetlands to open water has ranged from a low of 25 square miles to a high of 60 square miles *per year (*see Kelley et al., 1984, and Louisiana Coastal Wetlands Conservation and Restoration Task Force, 1993). Elsewhere in the United States, thousands of free-flowing streams have been straightened, channelized, or converted to slack water behind flood control dams (L. R. Johnston and Associates, 1991, Chapter 12). Along the coast, hundreds of miles of shoreline have been "hardened" to protect adjacent property from hurricanes and coastal storms so that, in cases such as southern New Jersey and New Hampshire, solid walls of concrete and rip-rap now line the shore rather than dunes and dry-sand beach, which have all but disappeared (see Beatley et al., 1994).

LAND USE MANAGEMENT—THE NEWEST APPROACH

The various shortcomings of warning and education, amelioration through relief and insurance, and hazard reduction through structural protection were well recognized by 1950. In that year, President Truman's Water Resources Policy Commission recommended that federal agencies work to foster a new approach—land use adjustments through zoning and other measures—as a way of reducing flood losses. The new land use approach the president's commission advocated had a simple premise: flood losses could be reduced significantly if, rather than trying to keep the flood out of people's way, government worked to keep people out of the flood's way by discouraging development of hazardous areas or, where development is warranted on economic grounds and little environmental harm results, by imposing special building standards that reduce vulnerability. These two types of land use measures, location and design, are still used today. (Land use approaches can mitigate the effects of an array of natural hazards, as described in Sidebar 1-1.)

The goal of the locational approach is to reduce losses in future disasters by *limiting development in hazardous areas.* This approach tends to be effective in reducing losses, preserving environmental values, and providing opportunities for outdoor recreation. But these gains come at the cost of giving up some of the economic benefits offered by hazard-prone land. The goal of the design approach, in contrast, is *safe con-*

struction in hazardous areas. This type of land use allows economic gains to be realized, but at the cost of greater loss of natural values and of susceptibility to greater damage when events overpower the design standards employed. A properly conducted planning process allows communities to find the right mix of these two land use approaches.

In managing the location of development, local governments work to shift existing development and steer new development to areas that are relatively hazard free. The tools communities can use for these purposes include land use regulations such as zoning, and a number of nonregulatory techniques, such as buying hazardous areas and using them for parks, buying repeatedly damaged structures and helping people relocate to a safer area, and locating development-inducing public facilities and services only in areas that are relatively free of hazards.

In managing the design of development, communities also can employ regulatory and nonregulatory techniques. Regulatory techniques include building codes and stand-alone ordinances that require actions such as elevating structures above estimated flood levels, bracing buildings to minimize damage from ground shaking during earthquakes, and

SIDEBAR 1-1

Hazards and Land Use

Coastal Storms and Hurricanes: Coastal storms such as northeasters damage beaches and waterfront buildings through shoreline erosion and flooding. Hurricanes add two additional threats: destructive winds of 75 to 200 miles per hour and wind-driven storm surges of up to 20 feet above sea level. Most hurricane-related deaths and property damage occur as a result of storm surges, but high winds and flying debris also can be enormously destructive. Other losses result from still-water flooding and tornadoes spawned by hurricanes. Hurricane damage tends to be concentrated in a narrow strip along the Atlantic Ocean and the Gulf of Mexico, but occasionally storms move inland and cause extensive flooding and wind damage as well. Because areas at highest risk from hurricanes are known, land use plans and management techniques can be used to good effect. Land use management approaches include limiting the intensity of development in areas subject to the highest storm surges, setting structures back from the shoreline to minimize losses from erosion, elevating structures to minimize the risk of flood damage, taking steps to tie buildings together (roofs to walls and installing hurricane clips so that roofs stay attached to buildings in high winds. Nonregulatory techniques include public information and training programs to inform builders and building owners of damage-resistant design techniques, and low-cost loans and other types of subsidies that make the new designs affordable. In the almost half-century since the federal government first began espousing land use approaches to hazard reduction, literally dozens of manuals and other training tools have been prepared to show local governments and property owners how to apply locational and design techniques.

Land use design measures are effective in reducing losses to new development but have less impact on losses to existing development in hazard zones. The reasons are economic and political. In the case of new construction, developers can take into account the higher costs imposed by more stringent design standards before they begin a project and can work these costs into their calculations, passing them backward to landowners through lower offering prices for development sites or forward to final consumers. Astute landowners will oppose land use regulations that affect new development, but developers and builders are

---

walls to foundations) so that they can withstand high winds, and protecting building openings such as doors and windows from penetration by flying debris. Losses in hurricanes also can be reduced by various engineering techniques, such as erecting sea walls, dikes, groins, and similar structures to hold back the ocean and minimize shoreline erosion.

Earthquakes: Losses occur in earthquakes as a result of ground shaking and from related hazards such as ground failures (from surface faulting, soil liquefaction, and landslides), flooding (from tsunamis and dam failures), and fire (as was the case with the San Francisco earthquake in 1906). Most of the United States is at risk from earthquakes, even though the most severe events over the past two centuries have been experienced along the Pacific Coast, in the Rocky Mountains, in the New Madrid region of Missouri, and in Charleston, South Carolina. Earthquake losses can be reduced if land use management techniques are employed to limit development in areas subject to ground failure and to require that buildings be able to withstand shaking. But the effectiveness of these techniques is limited because of difficulties in identifying faults, predicting the degree of shaking that is likely in small areas, and estimating the likelihood that related hazards

continued

likely to be neutral or supportive if development opportunities are not wholly foreclosed. In the case of existing construction, however, the added costs of retrofitting to meet more stringent design standards are not likely to have been anticipated when homes and other buildings were acquired and are hard to shift to someone else. As a result, researchers have found that imposing new standards on existing buildings through regulation is politically contentious and usually not very effective (for evidence of this, see Alesch and Petak, 1986, and Wyner and Mann, 1986). For similar reasons, trying to persuade owners to voluntarily retrofit their buildings to reduce the chances of damage in a disaster in most cases also has proven futile (see Laska, 1991). However, if information and regulations are supplemented with economic incentives, they have some potential to bring about relocation and retrofitting, as shown by various successful community efforts following the 1993 Midwest floods.

With the right mix of regulatory and nonregulatory land use management measures, local governments can achieve striking reductions in losses from natural disasters while at the same time accomplishing envi-

SIDEBAR 1-1 *Continued*

will be triggered. Also, as with other hazards, it is difficult to retrofit buildings once they have been constructed in an inappropriate location or using an inadequate structural design.

Floods: As a result of a variety of events—overflowing streams and rivers, melting snow, ice jams, dam and levee failures, or heavy rains coupled with inadequate drainage systems—more than 20,000 U.S. communities have problems with flooding. In fact, every year in the United States flooding accounts for the largest proportion of property losses resulting from natural disasters. Warning systems now avert most losses of life in flooding, but flash floods where little advance warning is possible are still a threat to life in a number of regions. But flood hazard areas can be identified before flooding occurs, and flood losses can be averted by avoiding building in such areas in the first place or by building in areas at risk but elevating structures above expected flood heights. Land use management techniques such as relocation and building elevation and floodproofing can be used to deal with existing development located in flood hazard areas, but costs tend to be high, and property owners often are reluctant to make changes to obtain the uncertain future benefits of reduced flood damages.

ronmental and other community goals. Burby and his associates conducted field studies to see what a national sample of ten cities had accomplished through planning and managing flood hazard areas over a ten-year period. The ten cities included: Arvada, Colorado (a suburb of Denver); Cape Girardeau, Missouri; Fargo, North Dakota; Omaha, Nebraska; Palatine, Illinois (a suburb of Chicago); Savannah, Georgia; Scottsdale, Arizona (a suburb of Phoenix); Toledo, Ohio; Tulsa, Oklahoma; and Wayne Township in northern New Jersey. The data collected in these cities revealed that over the ten-year study period, floodplain development had been reduced by over 75 percent of what would have occurred without the local planning programs (see Burby et al., 1988). Comparison of the benefits and costs of managing development ($11 million per year in reduced property damages versus $1.3 million in administrative and private costs) showed substantial net benefits from the efforts of these cities. But this and other studies have uncovered a number of serious barriers that have limited wider realization of these potential gains.

Landslides: Losses from landslides are widely distributed across the United States and are much lower than from the hazards listed above. Nevertheless, losses are estimated in the hundreds of millions of dollars annually, and they are apparently rising as development in urban areas expands onto marginal land or seeks to take advantage of views available from hillsides. Losses from landslides can be reduced if slide-prone areas are identified and development is limited. Even when it is difficult to predict the likelihood of slides with any degree of accuracy, development management techniques can be used to good advantage. For example, a number of communities have strict controls on grading of hillsides and require geologic studies and appropriate mitigation measures whenever development occurs on steep slopes. In addition, the likelihood of slides can be reduced through appropriate site-design measures such as retaining walls, drainage systems, and limitations on watering that may overload the bearing capacity of slope soils. After slides begin, corrective engineering measures can be taken, but they tend to be enormously expensive.

Wildfire: Loss of homes to wildfires has increased dramatically in recent years as urban growth has extended outward into what is termed the urban-wildland

continued

CRITICAL BARRIERS

As with the earlier approaches to hazard mitigation, land use management has had to contend with some serious problems that have limited its use and effectiveness. The most important of these problems are a lack of local political will to manage land use, deficiencies in management capacity, private-sector evasion of development rules, and regional fragmentation that limits opportunities for areawide management solutions. Showing how such barriers can be overcome is an important goal of this book.

A Crisis of Commitment

One of the most serious limitations of the land use approach is that without strong mandates from higher-level governments, few local governments are willing to protect against natural hazards by managing development. As noted earlier, in some cases land use measures are not appropriate. More often, however, local governments simply ignore the

SIDEBAR 1-1 *Continued*

interface. Combustible homes located in close proximity to combustible forests and other vegetation, in combination with minimal fire suppression capacity and inadequate water supplies for fire fighting, all contribute to fire losses. Land use measures can help prevent these losses if dense development is limited in areas with high fire hazards, if fire-resistant site design (for example, use of appropriate fuel breaks) and exterior building materials are required, if alternative ingress-egress routes are provided so that residents can escape if a primary route is blocked, and if adequate water supplies for fire suppression are required as a condition for development above certain densities.

Other Hazards: Land use methods can mitigate or eliminate the impacts of a number of other hazards, such as avalanches, drought, tsunamis, unstable soils, and volcanic mud and lava flows. However, land use measures are of little help with some events because locations most susceptible to risk cannot be located accurately enough or adequately differentiated from areas of lower risk. Examples of these hazards include frost and freezing, hail, lightning, snow, tornadoes, and windstorms (other than hurricanes).

Wave impact on houses in Scituate, Massachusetts, after a nor'easter in 1991 (Ralph Crossen, Building Commissioner, Barnstable, Mass.).

potential for losses. Like individuals, they tend to view natural hazards as a minor problem that can take a back seat to more pressing local concerns such as unemployment, crime, housing affordability, and education. One study, for example, found that on average local officials rank natural hazards thirteenth in importance, just behind pornographic literature, among the issues with which they are dealing (Rossi et al., 1982). Local officials become most interested in land use management when their communities suffer chronic losses. But by then planning is much less effective. Planning is, by definition, preventive in nature and not well suited to correcting problems with hazards once they have been allowed to develop.

Land use management was first advocated by the federal government in the 1950s, but by the end of that decade less than 100 of some 20,000 flood-prone communities were using zoning to limit more intensive use of flood hazard areas (Murphy, 1958). By the end of the 1960s, less than 500 had adopted such measures, and even after passage of the National Flood Insurance Act of 1968 (P.L. 90-448, Title XIII; *U.S. Code*, vol. 42, sec. 4201, as amended, 1968), which mandated building regulation as a condition for community eligibility for federally subsidized flood insurance, only 3,000 took part in the federal flood insurance program. Widespread participation in the program and adoption of the required building regulations did not come until after passage of the Flood Disaster Protection Act in 1973 (P.L. 93-234), which threatened to withdraw federal insurance for mortgages and funding of infrastructure (such as sanitary sewers) in flood-prone areas if communities did not adopt the requisite regulations. Today, about 90 percent of local governments with identified flood hazards participate and have adopted the federally mandated design standards. But for other hazards where a federal mandate does not require local regulation, much less has been accomplished.

Shortfalls in Management Capacity

The science of identifying hazards and designing to reduce their adverse impacts has far outrun the ability of local governments to put this new knowledge into practice. Land use approaches are founded on the accurate identification of areas affected by hazards, but hazard zone mapping is enormously expensive. In the case of flood hazards, for example, little progress was made in mapping areas subject to flooding until the 1950s, when the Tennessee Valley Authority, U.S. Army Corps of Engineers, and U.S. Geological Survey began helping with this effort. Even then, floodplain maps did not become widely available until the 1970s, when the federal government invested almost $1 billion in the production of maps needed to implement the National Flood Insurance Program. A similar federal effort has not been undertaken for other hazards, and local hazard identification often is sorely deficient.

Hazard zone mapping makes possible the adoption of both locational and design approaches to hazard mitigation. However, application of hazard-resistant methods requires knowledge and expertise often lacking in local government.

Few planning education programs provide detailed instruction in hazard mitigation and, as a result, few local planners have more than rudimentary knowledge of how to deal with hazards. Studies also have found that enforcement personnel may have insufficient knowledge of the hazard mitigation aspects of codes to enforce them effectively (see, for example, Southern Building Code Congress International, 1992). In addition, inadequate staffing may make it difficult for enforcement personnel, even if they are competent to interpret and apply code provisions, to keep up with the demand for building inspection (see Building Performance Assessment Team, 1992).

The Compliance Conundrum

A third critical barrier is lack of private sector compliance with land use and building regulations. Even if state and local governments adopt strict regulations to mitigate losses from natural disasters, the measures will not achieve their intended purpose if they are ignored, wholly or in part, by property owners, land developers, and builders. Massive breakdowns in enforcement and compliance with building codes have been implicated in the extraordinary losses experienced in recent disasters. In the case of Hurricane Andrew, for example, fully one fourth of insured losses were traced to shoddy design and construction practices. Analyses of building-damage distributions indicate that houses built after 1980 were 68 percent more likely to be uninhabitable after Andrew than older homes, even though building code standards applicable to newer houses were more stringent (*Miami Herald*, December 20, 1992, p. 3). Substandard construction practices were also reported as an important cause of heavy losses in Hurricane Hugo (All-Industry Research Advisory Council, 1989).

The Failure to Act Regionally

Yet another problem stems from the fact that hazards and the geophysical systems that engender them do not respect political boundaries. In some cases, land use management programs for hazard mitigation cannot be effective without intergovernmental coordination. Floodplain management experts Jon Kusler and Larry Larson (1993), for example, provide convincing evidence that the geographic scope of land use management needs to be expanded so that it embraces entire watersheds. When only a portion of a watershed is managed, serious consequences can result, such as when the hazard mitigation efforts of one jurisdiction lead to increased, not decreased, losses for neighboring communities. This sad result was documented for Jackson, Mississippi, which suffered millions of dollars in property damages over the Easter weekend of 1979 when flood waters of the Pearl River were pushed into the city by the levees of an adjacent city on the opposite bank of the river (see Platt, 1982). The multijurisdictional coordination needed to avert such disasters is very difficult to achieve. Even more difficult, however, is multijurisdictional agreement to management programs that have enough substance to actually accomplish a marked decrease in vulnerability.

Finding the Right Mix: The Role of Planning

The various barriers to effective land use management strongly suggest that this approach is likely to be ineffective if used as the sole means of averting losses in natural disasters, just as warning, relief, insurance, and hazard reduction would each be inadequate if used alone. The appropriate mix of these approaches is likely to vary from one place to another depending upon local circumstances. Because of this variation, it is virtually impossible for federal or state agencies to prescribe from above what communities should do to lessen their vulnerability and enhance sustainability. Instead, we believe *local planning* has to be conducted if communities are to find the right combination of measures that at the same time are effective, efficient, equitable, and feasible.

Planning for hazard mitigation should ensure that

- Information on the nature of possible future hazard events is available to the public.
- Land subject to natural hazards is identified and managed in a manner compatible with the type, assessed frequency, and damage potential of the hazard.
- Land subject to hazards is managed with due regard for the social, economic, aesthetic, and ecological costs and benefits to individuals as well as the community, while taking into account the rights of private landowners.
- All reasonable measures are taken to avoid hazards and potential damage to existing properties at risk.
- All reasonable measures are taken to alleviate the hazard and damage potential resulting from development in hazardous areas.

An example of a land use planning process is summarized in Sidebar 1-2. The detailed planning illustrated in the sidebar can be undertaken as the first step in a stand-alone hazard mitigation effort, or it can be incorporated as one element of a local comprehensive planning program. Less intensive planning for hazard mitigation also can be effective, if communities systematically consider hazards in the course of ongoing comprehensive and neighborhood planning.

Recent studies in Australia, New Zealand, and the United States have documented a number of benefits that follow when governments plan before they act (Burby and May et al., 1997; May et al., 1996). First, through the planning process, individuals and community policymakers can acquire information about the location and nature of various hazards, and about the liabilities of building in hazardous areas. Second, planning can yield information about the most appropriate uses of land in a community, showing that in many cases hazardous areas do not have to be used more intensively for communities to realize economic and other development objectives, and making it possible for communities to consider restrictions on building in hazardous areas, and to adopt them where economically efficient. Third, linking natural hazards to other, higher-priority issues confronting decision makers can lead to increased priority for hazard mitigation. Finally, the planning process, by involving all stakeholders in finding mutually beneficial ways of dealing with hazards, can help build a political constituency that will demand action to deal with hazards and will support appropriate hazard mitigation measures. Because little can be done if policymakers are unwilling to devote public resources to hazard reduction, the last benefit is particularly noteworthy and, by itself, probably justifies the time and effort required to plan.

CREATING NEEDED PARTNERSHIPS

Progress toward development and implementation of land use plans for hazard mitigation requires commitment to and capacity for hazard mitigation at all levels of government. Developing the necessary commitment and capacity requires that partnerships be formed horizontally at the community level and vertically with higher levels of government. The needed horizontal partnerships can be forged by a hazard mitigation committee and plan as described in Sidebar 1-2. The key vertical linkages that are essential to land use planning for hazard mitigation require attention by the states and the federal government.

The states and the federal government must play a lead role in building local capacity for and commitment to land use planning for hazard mitigation. The capacity of local governments can be enhanced by developing and providing information about natural hazards and by conducting research to formulate model hazard mitigation codes and standards. Training is equally important. Hazard mitigation is a new concept to many personnel in local government. Hazard mitigation training programs could include instruction in each of the components of mitigation planning, together with case studies showing how mitigation measures have been used in different types of communities.

But capability, once developed, is of no use without the willingness to use it to deal with natural hazards. Here higher levels of government have two important roles to play. The first is to lead by example. Governments at all levels have to take care that new facilities they fund are located, designed, and built in ways that minimize exposure to loss from natural hazards. The president issued executive orders in 1977 (floods) and 1990 (earthquakes) to direct federal agencies to avoid hazardous areas when at all practical and to construct all new buildings to be hazard resistant. The states need to take similar actions. The second and by far most important role of national and state governments, however, is to require, not just encourage, lower-level governments to prepare and implement land use plans for hazard mitigation.

If local governments are to prepare and follow through with land use planning for hazard mitigation, experience suggests that nothing short of strong mandates backed by the commitment and effort of the

SIDEBAR 1-2

A Five-Step Planning Process

Step One: Establish a Hazard Mitigation Committee

The first step is to form a committee including representatives of higher-level government who have expertise in hazard mitigation and representatives of all of the local groups who have a stake in mitigation. Typically these would include appropriate staff of the local government (building inspection, emergency management, engineering/public works, community planning) and representatives of economic interests, property owners, neighborhood groups, environmental groups, and real estate agents, bankers, developers, builders, and others involved in the land development process. The principal role of the committee is to assist the local governing body in developing and implementing a land use plan for hazard mitigation that enhances the sustainability of the community. In addition, the committee assists with

- directing and integrating supporting studies made at various stages of the planning process;
- formulating and applying interim development controls for use until the plan has been completed, approved, and implemented; and.
- developing strategies for implementing the plan.

state and federal governments are required. Without them, most local governments will continue with business as usual. A few innovative jurisdictions—those with extraordinary local leadership and those that have suffered severe losses from hazards in the past—will plan for and manage land use in hazardous areas. Most, however, will not, either because they lack adequate information about hazards and planning, or, more importantly, because there is no local constituency pushing in this direction. Thus, hazard mitigation requires a partnership. Impetus for land use planning and management must come from above, but the actual planning and conduct of programs must occur at the local level.

ABOUT THIS BOOK

Hazard mitigation—actions taken to prevent and reduce risks to life, property, and social and economic activities—is a key focus of the Inter-

---

A key function of the committee is to ensure that all relevant interests are involved in the planning process, and that there is some degree of consensus among affected groups that the plan is appropriate and will advance the interests of all concerned parties.

Step Two: Conduct Hazard Assessment Studies

The second step involves identifying and developing information about the hazards that threaten the community and their likely effects (e.g., the number of people who could suffer losses, and potential damages and other economic consequences). In the case of flood hazards, for example, the hazard assessment would identify likely flood discharges for rainstorms with varying recurrence intervals (hydrologic aspects), water levels and velocities (hydraulic aspects), and average annual flood damages to existing and possible future development (economic and social aspects). The sophistication and degree of detail of the study will vary depending upon the population and amount of property at risk. In highly urbanized areas, hazard assessments are prepared with the assistance of consulting engineers and may be rather expensive. In rural areas and small towns, more rudimentary studies, based largely on information from past occurrences of particular hazards, may be adequate to delineate the area at risk and possible conse-

continued

national Decade for Natural Disaster Reduction. This introductory chapter has described past responses to the threat of natural disasters and has outlined an approach to hazard mitigation based on local planning and land use management. In the remaining chapters of this book, a team of planning scholars expand on the ideas introduced here and illustrate how they can be realized.

The chapters of Part One, "The Choices of the Past," describe the evolution of land use adjustments to natural hazards and diagnose causes of the current failed intergovernmental system. In Chapter 2, "Planning and Land Use Adjustments in Historical Perspective," geographer and lawyer Rutherford Platt chronicles the long struggle of societies around the world to come to grips with the forces of nature. He shows that many of the key ingredients of a land use approach to hazard mitigation have been known and experimented with for literally hundreds of years, but a variety of obstacles have limited their use—the most recent being

SIDEBAR 1-2 *Continued*

quences of extreme events. Based on these studies, however, communities should: delineate areas at risk; develop inventories of the number, types, and, for flood hazards, elevations of buildings concerned; identify critical facilities (hospitals, fire stations, chemical storage companies) located in hazardous areas; determine if hazardous areas also provide natural and beneficial functions (e.g., wetlands, riparian areas, and habitat for rare and endangered species); identify development trends and pressures on the areas at risk; and finally identify community goals for the areas identified as hazardous.

Step Three: Conduct Hazard Mitigation Studies

The third step is to identify and analyze options for mitigation of the hazards that have been identified in Step Two. At this point, the hazard mitigation committee and planning staff: (1) identify the institutions whose actions can affect the nature of the hazard, the extent of development at risk, and the use of various hazard mitigation measures; (2) identify community goals and objectives related to land use and hazard mitigation; and (3) identify potential components of a hazard mitigation program and specific measures that are appropriate for the community. In evaluating various measures, the hazard mitigation study considers physical factors that affect their suitability to the locality, economic factors

the growing property rights movement. Nevertheless, in the 1988 Stafford Act and the 1994–1995 effort to "reinvent" the Federal Emergency Management Agency, mitigation has been identified as the single most important means of reversing escalating disaster losses (Robert T. Stafford Disaster Relief and Emergency Act, P.L. 100-707, 42 *U.S. Code* sec. 5121 et seq.). Thus, after repeatedly coming to the well of land use planning for mitigation but not actually drinking, federal leadership now may be ready to help adopt the approach we advocate.

The job of Chapter 3, "Governing Land Use in Hazardous Areas with a Patchwork System" by political scientist Peter May and planning educator Robert Deyle, is to explain why governance systems have failed to prevent the breakdown in sustainability that disasters signal. Likening the nation's approach to hazard mitigation to an old house that has undergone repeated remodeling and cosmetic "fixes" to meet the needs of new owners, they conclude that the resulting patchwork "system" is

---

such as costs in relation to benefits obtained from specific measures, and ecological factors that may be enhanced or adversely affected by various mitigation options. In addition, in collaboration with affected interests, the mitigation study should examine key issues related to past and future use of areas at risk. Issues may revolve around degrees of risk that are acceptable to the community, trade-offs to be made between economic development versus protection of aesthetic and environmental values, and the extent to which the use of private property can be restricted without compensating affected landowners. The hazard mitigation studies undertaken should consider all feasible mitigation options and should highlight all important issues related to the use of each option as it affects present and future land use.

Step Four: Prepare a Plan

Next, a land use plan for hazard mitigation is prepared for subsequent adoption by the governing body. Key components of the plan are:

- description of the plan objectives;
- discussion of the issues, problems, special features, and values specific to the areas covered by the plan;

continued

awkward, inconsistent in design, and not functioning well. By examining each of the system's component levels of government, however, they are able to isolate critical weak points and point toward needed reforms.

The chapters of Part Two, "The Land Use Planning Alternative," present four key components of successful planning for hazards—plans, information, development management, and partnerships with the private sector. In Chapter 4, "Integrating Hazard Mitigation and Local Land Use Planning," land use and environmental planning experts David Godschalk, Edward Kaiser, and Philip Berke share their experience in crafting land use plans for hazard mitigation. They describe the planning functions governments must engage in and illustrate what the resulting plans should look like. Because planning is not a static activity, they also discuss how plans can be evaluated so that planners learn from their experience and improve future planning products and processes.

In Chapter 5, "Hazard Assessment: The Factual Basis for Planning

SIDEBAR 1-2 *Continued*

- discussion of hazard mitigation policies and, possibly, a schedule of specific hazard mitigation measures to be undertaken;
- description of how hazardous areas are to be used and managed over the next 10 to 20 years;
- description of the means and timing of implementation, including designation of responsible individuals and agencies, specification of costs and financing, and specification of any necessary legislative changes; and
- discussion of approaches to monitoring the implementation and impacts of the plan and specification of procedures for periodically updating the plan (e.g., every five to seven years).

The primary objectives of the plan are to ensure that existing development and future development are compatible with the hazards that have been identified, and to identify hazard reduction measures that help ensure that land use in hazardous areas will be effective over the long term. In addition, it is critical that the plan contain an action agenda consisting of measures that can be undertaken in the short term (over the next 12 months) so that the plan serves as a living document and not an obscure blueprint that is put on the shelf and soon forgotten. The plan may be prepared as a stand-alone guide to local decision making

and Management," planning educators Robert Deyle, Steven French, Robert Olshansky, and Robert Paterson explain why natural hazard information and hazard assessment are an indispensable part of a land use planning approach. They describe how hazard assessment has been and can be applied to land use planning and management. Their analysis highlights advances and limitations of these methods and the choices that must be made if hazard assessment is to be effective in supporting mitigation.

Chapter 6, "Managing Land Use to Build Resilience," by planning educators Robert Olshansky and Jack Kartez, examines the various tools governments can use to shape land development and redevelopment decisions. The authors also review what has been learned over the past 20 years about the willingness of local governments to actually employ various measures to reduce risk. Obviously, accurate information and good plans mean little if they are ignored by policymakers. The authors show

---

about land use and hazards, or as one element of a broader comprehensive plan for community development.

Step Five: Implement the Plan

Once a plan has been adopted by the governing body, the hazard mitigation committee is not disbanded, but continues to meet regularly to monitor progress in accomplishing the measures that have been specified. Land use plans usually take years to accomplish in their entirety. The hazard mitigation committee is essential to keeping the constituency for mitigation intact and to keeping the important task of mitigation on the agenda of local and higher-level governmental leaders.

Source: The process described here is based on the "merits" floodplain management planning process developed by the State of New South Wales in Australia (see Department of Public Works, Government of New South Wales, 1986 and 1990).

that policy innovation and adoption follow a fairly regular course in localities that are pursuing aggressive land use planning programs. The analysis highlights the various elements that ease adoption, separating those that are external to government from those that are internal and can serve as policy targets.

Planning, hazard assessment, and land use management can take communities a long way toward increasing sustainability by reducing risk from natural hazards. In addition, however, governments need the capacity to actually undertake the new tasks the land use approach demands. Chapter 7, "The Third Sector: Evolving Partnerships in Hazard Mitigation," by planning educator Robert Paterson discusses creating this capacity through partnerships between nongovernmental groups. Paterson argues that nongovernmental organizations and private voluntary organizations can play a number of roles to make mitigation work successfully. In arguing this case, he provides examples from all parts of the nation of new voluntary efforts that address a variety of hazards.

The two chapters of Part Three, "Looking to the Future," lay out our conception of sustainability as a goal for land use planning and management, and then examine what the federal government can do differently to move society toward more sustainable urban development by mitigating hazards. In Chapter 8, "The Vision of Sustainable Communities," planning educator Timothy Beatley explains why sustainability is critical to the welfare of the country and shows how it can guide decisions about mitigation. Importantly, Beatley demonstrates that sustainability does not mean non-use of areas at risk from hazards, but *balance* of the important economic, environmental, and social interests at play there. We conclude our exploration of the land use approach to disaster reduction in Chapter 9, "Policy Directions for the States and Nation" by planning educator Raymond Burby, with contributions from each of the other chapter authors. Here we present five principles to guide the nation toward our vision of sustainable communities. Using these principles as a standard, we discuss the accomplishments and shortfalls in federal and state policy from four perspectives (objectives, approaches, governance, and scope) and then look at each principle and identify reforms in federal and state policies and programs that could foster the sustainable future we envision. These include policy changes involving incremental alteration of current programs, presidential and gubernatorial executive orders, and major new legislation. We conclude with suggestions for research to improve the performance of existing hazard mitigation programs and to facilitate policy reform.

PART ONE

The Choices of the Past

CHAPTER TWO

Planning and Land Use Adjustments in Historical Perspective

RUTHERFORD H. PLATT

FROM THE DAWN OF CIVILIZATION, humans have struggled to predict, prepare for, survive, and recover from natural disasters. The tale of Noah's flood in the Old Testament Book of Genesis, whether or not apocryphal, represents perhaps the earliest "documented" instance of flood hazard mitigation. One can assume that antiquity witnessed countless unrecorded episodes of coping with such eternal perils as fire, flood, earthquake, volcano, drought, famine, plague, and pestilence. The civilizations of the ancient Nile and Tigris-Euphrates Valleys clearly recognized how to derive agricultural benefits from floodplains while placing their settlements on higher ground if available. The colonial enclaves established by Greece and Rome chose sites, where possible, that were both militarily defensible and relatively safe from flood.

But the historical record on natural hazard mitigation, at least in Western Civilization, is sketchy until the Renaissance. Advances in literacy, technology, scientific knowledge, and the diminishing influence of unquestioning belief in an omnipotent and punishing God all contributed to the awakening recognition that human settlements may be designed or redesigned to be safer from natural and other perils. In fact, the idea that ordinary people are worth protecting from disaster was a precondition to examining why and where disasters occur and what may be done to reduce their effects.

LEARNING FROM DISASTER

The beginning of wisdom concerning disasters is the recognition that choices must be made, both in the original layout of new settlements in hazardous areas, and in recovery after a disaster strikes. Indeed, the aftermath of disaster offers a critical opportunity to review the choices available and possibly to select modification of the *status quo ante* to achieve greater protection from future disasters. Communities need not suffer destruction over and over again from the same natural hazard. As the experiences to be related from the Great Fire of London in 1666, the Lisbon earthquake of 1755, the San Francisco earthquake of 1906, and the Loma Prieta earthquake of 1989 illustrate, the recognition of choices in the wake of disaster is fundamental to greater security.

The Great Fire of London: Seizing the "Window of Opportunity"

Perhaps the first disaster to stimulate deliberate and well-documented changes in public policy was the Great Fire of London of 1666. This event, which burned most of the medieval part of the city within the old Roman Walls in three days, was faithfully chronicled by educated observers including Samuel Pepys and John Evelyn.

The fire epitomized Garrett Hardin's (1968) gloomy axiom "Freedom in a commons brings ruin to all." Neglect of the urban environment after a period of rapid population growth during the sixteenth and seventeenth centuries—combined with dry, windy conditions—led to disaster. The city's few public spaces had been encroached on by private structures clogging the narrow lanes and passageways and blocking access to the Thames River. In the absence of effective regulation of building size, location, and construction materials, the fire was inevitable. Without access to water, it could not be halted.

The point was not missed by certain leading minds of the time. While the ruins were still smoking, plans, or at least concepts, for the rebuilding of London were being prepared by Sir Christopher Wren, the city's leading architect, and several others (Bell [1920] 1971, Chapter 13).

Wren proposed to transform the city into a monumental, baroque-style imperial capital, much as Haussmann would restructure Paris in the mid-nineteenth century. But such a radical proposal for restructuring the city was incompatible with the culture and economy of England at that time. The government could not afford to pay property owners whose private lots would have been taken to implement the plan (a very modern dilemma).

A week after the fire had subsided, the recently restored sovereign, Charles II, addressed by proclamation the need for restraint and foresight in rebuilding, pending a full investigation of the causes of the disaster. Rasmussen ([1934] 1967, p. 117) praised this proclamation as "astonishingly modern . . . a true piece of town-planning, a programme for the development of the town." The proclamation addressed five practical aspects of the rebuilding process: (1) using stone or brick on exterior facades in place of wood; (2) setting minimum width of streets; (3) reserving open space along the Thames River to preserve access to water for firefighting; (4) removing public nuisances such as breweries or tanneries from central London; and (5) paying reasonable compensation to property owners whose right to rebuild was curtailed by public restrictions.

Charles II then appointed a special investigative commission (like a modern mayor appointing a "blue ribbon committee") to draw up recommendations for future building practices in London. Christopher Wren served on the commission and was to be a principal figure in the reconstruction of the city. Based on the royal proclamation and the commission's report, the "Act for Rebuilding London" was adopted February 8, 1667. The act has been described as London's first "complete code of building regulations" (Bell [1920] 1971, p. 251). While not fulfilling Wren's baroque vision (which he did instill in his new St. Paul's Cathedral), the rebuilt London proved durable: even the Nazi blitz failed to destroy London as completely as the Great Fire.

The Lisbon Earthquake of 1755: Assessing Options for Recovery

Less than a century after London's Fire, Portugal's capital of Lisbon was devastated by a major earthquake, which killed up to 30,000 people and destroyed much of its medieval core. As in England, experts were assembled under sovereign decree to assess whether and how the city should be rebuilt. The Marques de Pombal, a key royal advisor, was appointed to direct the recovery process. Recognizing the recurrent nature of the earthquake hazard, Pombal in modern fashion reviewed a range of options: (1) no change in streets or building lines; (2) widening of streets with little change in densities; (3) total demolition and replacement in the same location; (4) relocating the capital elsewhere (Mullin, 1992).

While London effectively followed the second approach, Lisbon chose the third. The Baixa district (the "lower town"), situated in the area of greatest seismic risk, was to be entirely leveled, replanned, and reconstructed. Pombal obtained a series of alternative plans for the rebuilding of the Baixa. The one selected accomplished several objectives: (1) it replaced the old medieval wooden city with stone and stucco in the manner of London, (2) it reoriented streets to gain maximum exposure to sunlight, (3) it linked commercial and royal districts of the city, (4) it eased circulation in the heart of the city, and (5) it created open space for ceremonies and access to the river. According to Mullin (1992, p. 168): "At once, ceremony, iconography, commerce, bureaucratic functions and everyday interaction were served."

As with the rebuilding of Paris by Baron Haussmann a century later, the reconstruction of Lisbon reflected the declining dominance of royalty and the church, and the emerging role of the commercial middle class. It provided order, relative safety from future earthquakes and fires, and an ornate capital setting for the Portuguese nation-state of the nineteenth century.

The San Francisco Earthquake of 1906: Confronting the Limits of Technology

It would be a century and a half before another catastrophic earthquake struck a large city: San Francisco in 1906. By the turn of the twentieth century, San Francisco had acquired a population of one-third of a million, and was well established as one of the leading American cities. Other cities, most recently Chicago, had experienced catastrophic fires. But no city in North America had been consumed by fire triggered by an earthquake. Like the sinking of the *Titanic* in 1912, the San Francisco earthquake and fire of 1906 shattered conventional assumptions regarding the ability of modern technology to overcome natural perils. And just as the loss of the *Titanic* led to tightening of safety requirements for ocean vessels, the San Francisco earthquake in time would stimulate actions there and elsewhere to mitigate the severity of future disasters.

The San Francisco Earthquake of 1906. This row of buildings tilted when the ground beneath them slumped because of the intensity of ground shaking. This photo was taken before fire destroyed the entire block (note smoke in the sky) (National Oceanic and Atmospheric Administration, National Geophysical Data Center).

The San Francisco earthquake and fire of April 18, 1906, is still considered America's worst urban natural disaster and one of the world's worst urban conflagrations (a list which now includes the Oakland 1991 firestorm). It burned an area of 490 blocks covering about 3,000 acres—six times the area burned by the Great Fire of London in 1666 and half again as much territory as the 1871 Chicago fire. The San Francisco fire and earthquake killed approximately 500 people and destroyed the entire business district and the homes of three-fifths of the city's inhabitants (Bronson, 1986).

The 1906 catastrophe exhibited many characteristics of contemporary urban natural disasters: (1) multiple interrelated hazards (earthquake and fire); (2) failure of lifelines (water, communications, transportation) which caused secondary impacts; (3) widespread structural damage due to inadequate building standards and prevalent use of wood for smaller structures; (4) resulting homelessness and joblessness of much of the working-class population; but (5) a nurturing external society that assisted in the immediate aftermath to the disaster and to some extent in the longer-term recovery. Missing from the San Francisco catastrophe, as compared with modern disasters, was any significant assistance from the national government. Most help was rendered by states, cities, churches, and other voluntary donors.

The foremost public action in response to the catastrophe was the development of an external water supply derived from damming of the Hetch Hetchy Valley in the Sierra Nevada, some 150 miles east of the city. The proposal by the City of San Francisco to construct a dam within Yosemite National Park precipitated a ten-year controversy between advocates of wilderness preservation, headed by John Muir, and proponents of "wise use of natural resources" represented by Gifford Pinchot, Director of the U.S. Forest Service and advisor to President Theodore Roosevelt. The dam and reservoir were finally approved in 1913 and have since provided San Francisco with a high-quality source of drinking water. But the reliability of the system in the event of another major Bay Area earthquake remains in doubt. The Hetch Hetchy water pipes extend across the Hayward Fault in the East Bay and thus could rupture in an earthquake there, potentially leaving San Francisco once again deprived of its water supply (a problem shared by its East Bay neighbors).

San Francisco recovered rapidly from the 1906 catastrophe chiefly because the downtown business district was well insured. Total insurance payments amounted to $5.44 billion (1992 dollars), by far the worst urban fire insurance loss in American history. However, like London after its fire in 1666, San Francisco declined to alter its basic pattern of streets and land use patterns as it rebuilt. In particular, it ignored Daniel H. Burnham's "City Beautiful" plan for the redesign of the city that had been presented to civic leaders just before the fire: "Few cities ever found themselves demolished, with a ready-made plan for a new and grander city already drawn up, awaiting implementation, and with money pouring in to help realize the plan. San Francisco chose to ignore its Burnham Plan, and decided instead to build at a rate and manner which made the city not only less beautiful than was possible, but more dangerous. The rubble of the 1906 disaster was pushed into the Bay; buildings were built on it. Those buildings will be among the most vulnerable when the next earthquake comes" (Thomas and Witts, 1971, p. 274). This statement was prophetic: the city's Marina district, built on 1906 rubble, sustained heavy damage in the 1989 Loma Prieta earthquake.

The Loma Prieta Earthquake: Lifeline Failures

Modern urban disasters, particularly earthquakes, involve disruption of public and private lifelines such as bridges and highways, communications, electrical facilities, water, food supplies, and medical care. The failure of lifelines, by definition, amplifies the geographic, economic, and social consequences of a natural disaster. The Loma Prieta earthquake of 1989 was, like the 1906 San Francisco earthquake, notable for significant lifeline failures. Such failures on an even larger scale followed the Northridge earthquake of 1994 and the Kobe earthquake in Japan in 1995.

On October 17, 1989, the San Andreas Fault reawakened in the Bay Area with a 7.1 magnitude earthquake, the first of that magnitude since 1906. The fault ruptured over a distance of 25 miles at a depth of 12 miles below the surface; no surface faulting appeared (Bolin, 1993b; FEMA, 1991). Seismic shaking lasted 15 seconds, followed by numerous aftershocks. The event was felt over an area of about 400,000 square miles extending southward as far as Los Angeles and northward to the Oregon border. It caused 62 known deaths and 3,757 injuries, and left 12,000 people homeless. The quake resulted in over $6 billion in property damage and disrupted public transportation, utilities, and communications. Occurring just before a World Series game, ironically between the San Francisco Giants and the Oakland Athletics, the earthquake presented an unparalleled opportunity to raise public awareness of seismic hazard.

Although the epicenter was in a rural upland well south of the major population centers of the Bay Area, it nevertheless caused dramatic and costly damage to infrastructure and older private buildings on both sides of the Bay. In general, French (1990) has identified four classes of infrastructure damage caused by the Loma Prieta quake:

1. direct physical and economic damage to infrastructure systems themselves
2. diminished ability to carry out emergency response activities
3. inconvenience due to temporary service interruption
4. longer-term economic losses due to the length of time required to rebuild.

The most horrifying result of the earthquake was the collapse of a 1.5 mile double-deck section of a major commuter freeway constructed on bay mud in Oakland. The collapse trapped hundreds of cars and accounted for 41 of the 62 deaths attributable to the earthquake. For several days, the nation witnessed on television the heroic rescue of survivors from their vehicles trapped under the collapsed rubble of the upper deck. The last living victim was removed from his car after 90 hours. (Even more extensive freeway damage would occur in the January 17, 1994, Northridge earthquake in southern California.)

Another lifeline failure was the collapse of a 50-foot section of the upper deck of the San Francisco–Oakland Bay Bridge, the only direct vehicular link between San Francisco and the East Bay. Its closure for repairs forced the cross-Bay commuters to take circuitous routes far to the north or south, or to rely on the BART rapid transit system, which was not damaged. Cross-bay ferry service was also revived. In San Francisco, Interstate Highway 280 and the incomplete Embarcadero Freeway were also damaged. Elsewhere, many local roads and highways were blocked by landslides triggered by ground shaking.

Structural damage was concentrated in cities close to the epicenter, particularly Santa Cruz. But many older unreinforced masonry buildings were damaged in Oakland (including its City Hall) and in San Francisco. Damage in both cities was primarily caused by ground shaking on unconsolidated filled lands along the bay, as predicted by Thomas and Witts (1971). As in 1906, local water mains ruptured, leading to loss of water pressure for firefighting. A fireboat helped to pump bay water and local citizens organized bucket brigades to save the Marina District.

Elsewhere, damage was more related to soil propensity for ground shaking than to distance from the epicenter. Many critical facilities including several hospitals were damaged in locations where hazardous soil conditions could have been recognized. Stanford University in Palo Alto, which was severely hit by the 1906 earthquake, suffered $160 million in damage from Loma Prieta. Public schools were generally not badly damaged due to earthquake construction codes adopted after the 1933 Long Beach earthquake in southern California. The East Bay Municipal Utility District (EBMUD), which supplies water to 1.5 million people in the East Bay region, experienced some 200 local water main breaks, but its main supplies were not affected (FEMA, 1990).

Loma Prieta was a wake-up call to the Bay Area, which would be reinforced by the 1994 Northridge earthquake in Southern California and the Kobe earthquake in 1995. While building codes have been gradually upgraded in California and elsewhere to protect against earthquake damage to individual structures, the prospect of massive lifeline failures in metropolitan regions is daunting. Preparedness for the proverbial "Big One" is the highest-priority issue in natural hazard planning for FEMA and the State of California.

FLOOD DISASTERS: A RANGE OF ADJUSTMENTS

The Mississippi-Missouri River system is second in the world only to the Amazon River in main-stem length (3,799 miles) and ranks behind only the Amazon and the Congo Rivers in watershed area. But in contrast to most other great rivers, the Mississippi-Missouri system drains a fertile and largely inhabitable heartland.

The Mississippi-Missouri watershed, comprising about 41 percent of the land area of the conterminous United States, is a region of great climate variability. Average annual precipitation ranges from less than 20 inches in the upper Missouri basin in the west to more than 50 inches in the Ohio River basin in the east. High levels of river flow result from seasonal snowmelt in the northern portions of the basin and from storm systems that move across the region from west to east at any time of year. Tropical hurricanes emerging from the Gulf of Mexico in late summer and fall occasionally threaten the lower valley with heavy rainfall, wind, and coastal storm surges. All of these climatic phenomena contribute to recurrent and occasionally massive floods along the Mississippi and its tributaries.

More than any other river system, Mississippi floods have consistently attracted national attention and, when flooding causes widespread losses, national response. The policies of the United States towards flooding have to a great extent been forged in the incessant quest for a reasonable accommodation between human settlement and natural hazard in the Mississippi River valley.

Early plantations established along the river in the eighteenth century under French rule sought to protect themselves from floods and channel shifts by building levees on their respective lands. These individual private levees to control the river were largely ineffective against major floods because they were discontinuous and of uneven height and durability.

With the accession of the region by the United States through the Louisiana Purchase of 1803, local interests increasingly sought governmental assistance and protection. The Army Board of Engineers in 1825 was authorized to undertake waterway improvements, but for several decades these were intended primarily to aid navigation. In 1861, two army engineers, Abbot and Humphreys, proposed a system of levees along the Mississippi River to enhance navigation and also provide flood control. This approach, known as the "levees only" policy, was adopted. A series of further floods prompted Congress to establish the Mississippi River Commission, a joint military-civilian board that continues to oversee federal efforts to contain the lower Mississippi today. The commission attracted humorous skepticism from Mark Twain in his 1882 book *Life on the Mississippi.* Notwithstanding Mark Twain's doubts, the lower Mississippi was lined with federally designed and funded levees by the early twentieth century. Levees were also beginning to appear along segments of the upper Mississippi and its navigable tributaries.

Faithful to the "levees only" policy, flood control dams for upstream storage were not built on the Mississippi before the 1930s. Upstream storage was first used on a large scale in the Miami River basin in Ohio following disastrous floods there in 1913. The Miami River Conservancy District, organized under a state law passed the following year, constructed a series of five dams to impound flood runoff in reservoirs that remained empty when not needed for that purpose.

The great lower Mississippi River flood of 1927 prompted Congress to scrap the "levees only" policy and expand the range of engineering approaches to controlling the river. The 1927 flood reclaimed most of the alluvial valley from the Ohio River to the Gulf. Hundreds of levees were breached or overtopped. Some 20,000 square miles of land were inundated, 700,000 people were displaced, over 200 were killed, and 135,000 structures were damaged or destroyed.

Seldom has a policy change been so quickly and clearly adopted in the wake of a natural disaster. The lower Mississippi Flood Control Act of 1928 authorized a series of dams and flood storage projects, channel improvements, floodways, and other measures for the valley. According to two eminent hydrologists, Hoyt and Langbein, "Few natural events have had a more lasting impact on our engineering concepts, economic thought, and political policy in the field of floods. Prior to 1927, control of floods in the United States was considered largely a local responsibility. Soon after 1927 the control of floods became a national problem and a federal responsibility" (1955, p. 261).

After disastrous floods in the Ohio Valley and New England in 1935 and 1936, the Flood Control Act of 1936 shifted the focus of Army Corps of Engineers flood activities beyond the Mississippi Valley to the entire nation. The act authorized 218 new projects nationally, including several multiple-purpose projects. Federal funding was contingent upon nonfederal cost-sharing and upon a benefit-cost analysis that determined that a project would be cost-effective. But in the Flood Control Act of 1938, Congress assumed full federal funding of upstream flood control dams and reservoirs, while levees remained subject to local contribution.

During the 1930s and the New Deal, some of the world's largest and most admired multipurpose river development projects were completed. Foremost among these was the system of main-stem and tributary dams constructed by the Tennessee Valley Authority (TVA). The TVA was chartered by Congress in 1933 as a public corporation to focus federal resources upon an impoverished and environmentally stressed region, the drainage basin of the Tennessee River. The TVA is best known for its series of main-stem dams which harnessed the river for power, navigation, recreation, and flood control. But the TVA also developed pioneering programs in floodplain management, soil erosion management, reforestation, economic development, and improvement of housing, medical care, schools, and recreation. It proved to be an internationally important experiment in governmental resource management. Although no other similar agencies were established in the United States, the TVA demonstrated that it is possible to promote regional development through comprehensive river basin management (White, 1969). The concept of river basin planning was further promoted through the many reports of the National Resources Planning Board during the 1930s.

Federal flood control and multipurpose river development projects were undertaken on the Columbia, the Colorado, the Sacramento, and in the east in more limited form on the Connecticut, the Delaware, the Potomac, and elsewhere. Grand Coulee Dam on the Columbia and Hoover Dam on the Colorado remain two of the most impressive concrete dams in the world.

Between 1936 and 1952, Congress spent more the $11.1 billion for flood control, of which $10 billion was allocated to the Corps of Engineers (not adjusted for inflation). The cost of Corps flood control projects in the lower Mississippi River valley alone between 1928 and 1983 was estimated at $10 billion. After the 1993 floods on the upper Mississippi and Missouri rivers, the Army Corps of Engineers reported that all federally constructed facilities performed as designed and prevented $19.1 billion in flood losses (Interagency Floodplain Management Review Committee, 1994, p. 21).

The concept of floodplain management through land use planning and restrictions was relatively late to appear. In the early 1950s, the Tennessee Valley Authority initiated a floodplain mapping and information program under the direction of James E. Goddard. This would in turn stimulate the Army Corps of Engineers to initiate a wider flood information program around 1961, which continued until the primary function of mapping floodplains was assumed by the National Flood Insurance Program. However, communities were slow to translate improved information into land use regulations. In 1955 Hoyt and Langbein wrote, "Flood zoning, like almost all that is virtuous, has great verbal support, but almost nothing has been done about it" (p. 95).

Nonstructural Approaches to Flood Control

The National Flood Insurance Program

The geographic focus of flood policy attention shifted to the Atlantic and Gulf of Mexico coasts in the wake of a series of vicious hurricanes during the 1950s and 1960s. In 1965, Congress called for a study of flood insurance and other measures as alternatives to structural flood control and disaster assistance. Two ensuing reports, respectively authored by resource economist Marion Clawson and geographer Gilbert F. White, recommended that a national flood insurance program might be feasible if it contained requirements for land use controls and building standards to reduce future losses. Congress responded by passing the National Flood Insurance Act.

This act established the National Flood Insurance Program (NFIP), which has become the primary vehicle of federal flood policy. This program serves three interrelated congressional objectives. First, the NFIP seeks to reallocate a portion of the burden of flood losses to all occupants of flood hazard areas through the mechanism of insurance premiums. Second, it seeks to reduce steadily increasing flood losses by limiting additional development and investment at risk in floodplains. Third, the NFIP calls for mapping of flood hazard areas across the United States; the program has spent over $900 million for this purpose since the law was enacted. In order for insurance to be available to property owners, the community in which their land is located must regulate new and rebuilt structures in the flood hazard area according to standards set by the NFIP, particularly the elevation of the lowest floor of a structure above the estimated local "base flood" level. Currently over 18,000 municipalities and counties have adopted such regulations and are eligible for flood insurance coverage.

The 1973 Flood Disaster Protection Act, adopted after Tropical Storm Agnes devastated the Middle Atlantic states, made flood insurance purchase compulsory for anyone borrowing money from a federally related lender to purchase or develop property in identified flood hazard areas. This requirement helped boost participation in the National Flood Insurance Program to a level of about 3.3 million policies by 1996, with total coverage of over $300 billion.

Besides the National Flood Insurance Program, nonstructural flood loss reduction has included several other measures, for example, improved weather forecasting and warning systems, preparedness and evacuation planning (particularly in areas subject to hurricanes), floodproofing of existing structures, and relocation of certain structures in locations subject to chronic flooding.

Also as a condition of eligibility for flood insurance, the National Flood Insurance Act required land use planning for and management of identified flood hazard areas. This planning and management was to be a joint responsibility of the federal government, states, and local communities. The Federal Insurance Administration (until 1979 a unit of the Department of Housing and Urban Development and thereafter of the Federal Emergency Management Agency) was required to "develop comprehensive criteria designed to encourage, where necessary, the adoption of adequate state and local measures which . . . will: (1) constrict the development of land which is exposed to flood damage where appropriate, [and] (2) guide the development of proposed construction away from locations which are threatened by flood hazards" (P.L. 90-448, sec. 1361; *U.S. Code*, vol. 42, sec. 4102).

Floodplain Zoning

The Flood Disaster Protection Act of 1973 made land use management and participation in the NFIP prerequisites for federal financial assistance including disaster assistance. And in contrast to the use of "encourage" in Section 1361 of the 1968 law, the 1973 law stated the intent of Congress to "*require* states or local communities as a condition of future federal financial assistance, to participate in the flood insurance program and to adopt adequate flood plain ordinances with effective enforcement provisions consistent with federal standards to reduce or avoid future flood losses" (P.L. 93-234, sec. 2(b)(3); *U.S. Code*, vol. 42, sec. 4002(b)(3), emphasis added).

But uncertainty lingered as to whether floodplain zoning that precluded use of privately owned flood hazard areas was constitutional or not. Since its origins in the early twentieth century, the planning and zoning movement in the United States had taken little notice of natural hazards. Zoning—the division of a community's land area into districts of specified land uses and densities—had largely been employed to protect single-family neighborhoods from incompatible neighbors, and later to insulate suburban communities from unwanted forms of housing, commerce, and public facilities. Until the late 1960s, natural hazards had generally been assumed to be controllable through technology and were therefore not considered an impediment to development.

In the first comprehensive legal review of floodplain zoning, law professor Allison Dunham (1959) found few judicial cases on the subject. However, he urged that such measures should be held valid to protect (1) unwary individuals from investing or dwelling in hazardous locations, (2) riparian landowners from higher flood levels due to ill-considered encroachment on floodplains by their neighbors, and (3) the community from the costs of rescue and disaster assistance.

In 1972, the Massachusetts Supreme Judicial Court in *Turnpike Realty Co. v. Town of Dedham* (284 N.E.2d, at 891) at last provided a strong decision upholding a local floodplain zoning ordinance. The case involved a 61-acre parcel located in a wetland adjoining the Charles River in eastern Massachusetts which the plaintiff sought to fill and develop. The town refused permission to do so under its 1963 "flood plain district" law. The purposes of the law included: (1) protection of the groundwater table; (2) protection of public health and safety against floods; (3) avoidance of community costs due to unwise construction in wetlands and floodplains; and (4) conservation of "natural conditions, wild life, and open spaces for the education, recreation and general welfare of the public" (284 N.E.2d, at 894). The last objective raised the specter of "public benefit" which normally is a forbidden purpose of zoning. The Massachusetts court, however, accepted the first three purposes as valid and ruled that since "the by-law is fully supported by other valid considerations of public welfare," the conservation of natural conditions would be acceptable as well.

Although the plaintiff claimed that the site was only flooded due to improper operation of a floodgate, the court found the site to lie within the natural floodplain of the Charles River. It cited testimony as to recent severe flooding of the site. Relying on the Dunham (1959) rationale stated above, the court upheld the measure, declaring, "The general necessity of flood plain zoning to reduce the damage to life and property caused by flooding is unquestionable" (284 N.E.2d, at 899).

During the 1970s and the 1980s, state and local floodplain regulations were widely adopted in response to the National Flood Insurance Program and increasing public recognition of coastal and riverine flood hazards. Upon challenges to such laws, courts generally followed the *Turnpike Realty* rationale and upheld the constitutionality of floodplain regulations (Kusler, 1982). State floodplain managers funded by FEMA assisted elected officials and judges in understanding flood hazards. Their professional organization, the Association of State Floodplain Managers, has contributed through its conferences and publications to the acceptance of floodplain management, including land use regulations.

LAND USE REGULATION REGARDING OTHER HAZARDS

In contrast to floods and coastal hazards, there is no congressional mandate for land use planning and regulation in the case of seismic hazards, landslides, or urban-wildland fire risk areas. Whatever the shortcomings of its implementation, the NFIP has at least promoted the idea of regulating development in identified flood hazard areas. The Earthquake Hazards Reduction Act of 1977 (P.L. 95-124; *U.S. Code*, vol. 42, sec. 7701) established a program of research and technical assistance to states and local governments but stopped short of establishing a mandate for planning and regulation of land use and building practices. In 1993, a review by the former Federal Insurance Administrator George Bernstein found mitigation of earthquake hazards to be inadequate due to the voluntary nature of the national program. In particular, he found that "neither improved design and construction methods and practices nor land use controls and redevelopment have been implemented broadly enough to hold out promise for comprehensive hazard reduction throughout the United States" (Bernstein, 1993, p. 28). As discussed below, land use planning has likewise not been utilized widely in the aftermath of such wildfire disasters as the Oakland fire of 1991.

THE PROPERTY RIGHTS MOVEMENT: PUTTING THE BRAKES ON LAND USE MANAGEMENT?

In the mid-1990s, the climate of political and judicial approval of floodplain regulations may be weakening. The "property rights movement" and the Republican "Contract with America" aroused broad grassroots opposition towards government regulations over the use of private land. Restrictions on building or rebuilding along hazardous coastal shorelines have been a particular area of controversy. Property rights organizations such as the Fire Island Association (New York) have argued strenuously that no limits on coastal building should be imposed without compensation to the private landowner. They and other property rights groups successfully blocked amendments to the National Flood Insurance Act that would have mandated minimum setbacks for new and rebuilt structures along eroding shorelines, as recommended by a 1990 report of the National Research Council (1990). After passing the House of Representatives by a vote of 388 to 18 in 1991, the measure never reached the Senate floor due to property rights lobbying. The 1994 NFIP amendments simply called for further study of coastal erosion (P.L. 103-325, sec. 577).

Property rights advocates have also sought to enlarge the scope of compensation for "regulatory takings" in several recent Supreme Court decisions. Their argument is based on the fifth amendment to the U.S. Constitution, which states in part: "nor shall private property be taken for public use without just compensation." Since the 1920s, it has been recognized that occasionally public regulations may "go too far" and amount to a "taking" of the value of property, for which compensation should be paid. Generally, where a regulation clearly is supported by a valid public purpose, such as preventing flood losses, it is upheld regardless of its impact on property values.

In 1987, the property rights movement gained a halfhearted decision from the U.S. Supreme Court relating to a flood disaster. *First English Lutheran Evangelical Church v. County of Los Angeles* (107 S.Ct., at 2378, 1987) involved a county moratorium on rebuilding of a camp for handicapped children in a canyon after a flash flood swept through the area. A 5-4 majority sustained a theory of "inverse condemnation" which allows an owner to recover monetary damages for loss of value during the time a restriction is in effect, if the restriction is held to be invalid. The Court did not decide whether the Los Angeles County moratorium was in fact a taking and remanded the question to the California Court of Appeals, which ruled that it was not and vigorously upheld the county's challenged moratorium: "If there is a hierarchy of interest the police power serves—and both logic and prior cases suggest there is—then the preservation of life must rank near the top. Zoning restrictions seldom serve public interests so far up on the scale. . . . The zoning regulation challenged in the instant case involves this highest of public interests—the prevention of death and injury. Its enactment was prompted by the loss of life in an earlier flood. And its avowed purpose is to prevent the loss of lives in future floods" (258 Cal. Rptr., at 904, 1989).

In 1992, the U.S. Supreme Court held in *Lucas v. South Carolina Coastal Council* (112 S.Ct., at 2886, 1992) that a regulation that removes virtually all of the value of property may require compensation as a "taking" *regardless of the purpose of the law.* Under the 1988 South Carolina Beachfront Management Act (Act 634), which prohibited new building seaward of an erosion setback baseline, Lucas was denied permission to build on his two oceanfront lots. Lucas did not challenge the validity of the Beachfront Management Act per se but claimed that its application to his lots destroyed all of their value. The trial court agreed and ordered the state to pay Lucas $1.2 million as compensation (Platt, 1992, 1996).

After the South Carolina Supreme Court (404 S.E.2d, at 895, 1991) reversed the trial court, the U.S. Supreme Court accepted Lucas's appeal from the state decision. The resulting national attention attracted numerous "amicus curiae" briefs by interested parties on both sides of the issue. Its potential importance was reflected in an editorial in the *Boston Globe* (March 5, 1992): "The case has far-reaching implications for the enforcement of regulations concerning everything from billboards to wetlands, as well as the coastline. Environmentalists fear that if the court decides in Lucas' favor, virtually every environmental restriction placed on the use of property will be considered a taking, thus making environmental protection too expensive" (p. 14).

The U.S. Supreme Court reversed the state ruling in a 6-3 decision, holding that where a regulation "denies all economically beneficial or productive use of land" (112 S.Ct., at 2893, 1992), it is a compensable taking. The majority opinion held that the need to compensate for "total takings" could not be avoided by merely reciting harms that the regulation would prevent, except in the case that a regulation merely reflected a state's "background principles of nuisance and property law" (Id., at 2901). Subsequently, the state supreme court on remand held that the Beachfront Management Act did not fall under that exception and that Lucas should be compensated for the period during which his land was restricted. (The state subsequently sold both of the Lucas lots to recoup its expenses, and a large house has been built on one of them, which in 1996 was threatened by renewed erosion!)

Dolan v. City of Tigard, Oregon (114 S.Ct., at 2309, 1994) involved a property owner's challenge to several requirements imposed by the city as conditions for approval of a permit to enlarge an existing hardware store. The owner was required to donate a portion of her property lying in the 100-year floodplain, plus an additional strip to be used as part of a public bikeway system. The U.S. Supreme Court reversed the Oregon Supreme Court in another 5-4 decision and held these conditions to be invalid takings due to a lack of "rough proportionality" between the burden upon the property owner and the benefit to the public: "No precise mathematical calculation is required, but the city must make some sort of individualized determination that the required dedication is related both in nature and extent to the impact of the proposed development" (114 S.Ct., at 2329–2330). The court was primarily interested in the demand by the city that public access by means of a bikeway be allowed by the property owner without compensation. The floodplain restriction was not discussed. It is not clear whether the court meant to apply the "rough proportionality" test to floodplain management regulations.

However, as with *Lucas*, the political importance of *Dolan* transcended its narrow legal significance. Justice Stevens observed in dissent that "rough proportionality" is difficult to satisfy in relation to such statistically uncertain phenomena as flooding and traffic congestion. In language calculated to raise the spirits of planners and floodplain managers, Stevens declared: "In our changing world one thing is certain: uncertainty will characterize predictions about the impact of new urban developments on the risks of floods, earthquakes, traffic congestion, or environmental harms. When there is doubt concerning the magnitude of those impacts, the public interest in averting them must outweigh the private interest of the commercial entrepreneur" (114 S.Ct., at 2329). With a change of one vote, this statement would have been the majority opinion and the law of the land.

In addition to challenges to land use regulation in the courts, property rights advocates have attempted to pass legislation that would permanently constrain takings by governmental action. Between 1990 and 1996, 18 states adopted takings legislation, while efforts to pass bills failed in 30 other states. These laws fall into two categories: assessment laws and compensation laws. An assessment law typically requires all governmental actions, or a select group of them, to be reviewed for potential impact on private property prior to implementation. The reviews, called "takings impact assessments," are in theory equivalent to the judicial review undertaken through inverse condemnation suits, but occur before rules and regulations are adopted and implemented. These laws are expected to have a "chilling" effect on more restrictive governmental actions in order to avoid invalidation. Compensation bills, which have been adopted by four states, go a step farther. If land use rules or regulations will diminish the fair market value of property by a specified percentage (usually ranging between 10 and 50 percent), property owners must be compensated before the rule or regulation takes effect (see Freilich and Doyle, 1994). Whether the rush to limit land use regulation through legislative means will continue or abate is uncertain.

DISASTER ASSISTANCE: CAN MITIGATION STEM THE TIDE?

Although floods are involved in about 90 percent of natural disasters in the United States, the National Flood Insurance Program has by no means succeeded in its goal of shifting the costs of floods from taxpayer-funded disaster assistance to premium-based flood insurance. Outlays for federal disaster assistance, which follow presidential disaster declarations, have risen from a paltry $5 million appropriated in the original 1950 Federal Disaster Relief Act to several billion dollars annually in the 1990s. Outlays from the Disaster Relief Fund accounted for 80 percent of the $5.4 billion in payments made by FEMA in fiscal year 1994 (U.S. Congress, General Accounting Office, 1995, p. 2). But outlays from the fund comprise only an indeterminate fraction of total federal disaster-related assistance provided each year. Approximately 30 federal programs of various types offer some form of disaster service or funding (Congressional Research Service, 1992, p. 11). Twenty-six federal departments and agencies perform disaster response functions (plus the American Red Cross) as part of the Federal Response Plan coordinated by FEMA (FEMA, 1992b).

Federal assistance to public and private disaster victims after a presidential declaration is largely in the form of grants funded by the nation's taxpayers. The Bipartisan Senate Task Force report on Federal Disaster Assistance estimated that federal disaster-related expenditures from 1977 through 1993—those devoted to preparedness, emergency response, recovery and mitigation (including structural flood control projects)—totaled about $119 billion (U.S. Senate Task Force on Funding Disaster Relief, 1995, p. viii). Of that amount, direct grants to communities and individual victims, including but not limited to the Disaster Assistance

Program, amounted to about $64 billion. The remainder, approximately $55 billion, occurred in the form of low-interest loans and insurance payments. For those outlays, the actual tax cost, after considering repayments of loans and payment of insurance premiums, is represented by the amount to which the federal government subsidizes insurance premiums, issues loans at subsidized rates, or forgives loans (i.e., does not require them to be repaid).

To stem the ever-rising tide of federal disaster assistance, hazard mitigation has been increasingly advocated. As discussed previously, hazard mitigation has long been employed on a national level to protect against floods. The 1974 federal Disaster Relief Act initiated a procedure for assessing all natural hazards following a disaster declaration in which "the state or local government shall agree that the natural hazards in the areas in which the proceeds of the grants or loans are to be used shall be evaluated and appropriate action shall be taken to mitigate such hazards, including safe land use and construction practices" (P.L. 93-488, sec. 406; renumbered sec. 409 in P.L. 100-707). The Office of Management and Budget in 1980 charged federal agencies under the leadership of FEMA to prepare post-disaster assessments of options for mitigating future flood losses. To comply with these mandates, FEMA formed interagency hazard mitigation teams in each federal region to prepare the required reports promptly after a disaster occurs.

The Stafford Disaster Relief and Emergency Assistance Act in 1988 required evaluation of mitigation opportunities to accompany efforts to recover from any disaster for which funds are expended under the act. Furthermore, a variety of initiatives such as the National Earthquake Hazard Reduction Program and FEMA's Hurricane Preparedness Program seek to promote pre-disaster planning and mitigation activities at the state, local, and private levels. FEMA in 1993 established a new Mitigation Directorate, which is formulating a "National Mitigation Strategy."

However, the recommendations of hazard mitigation assessments often fly in the face of the desire of disaster-stricken communities to rebuild as quickly as possible. Rebuilding more safely may be more expensive than simply reconstructing the *status quo ante*. Section 404 of the Stafford Act authorized federal funding of 50 percent of the cost of mitigation projects, up to a maximum of 10 percent of total Stafford Act funding under a particular disaster declaration. After the Midwest floods of 1993, this was amended by P.L. 103-181 to authorize 75 percent federal funding for mitigation, up to 15 percent of total disaster expenditures. A supplemental appropriations bill signed July 21, 1995, provided $6.5 billion for disaster relief but a mere $5 million for start-up of the Mitigation Assistance Grant Program.

Funding for flood-related mitigation projects, without the need for a disaster declaration, was authorized in the 1994 amendments to the National Flood Insurance Act (P.L. 103-325, Subtitle D). These amendments established a National Flood Mitigation Fund (NFMF), to be financed through surcharges on NFIP policies. Grants were authorized to states and communities for preparing "flood risk mitigation plans" such as those described in Chapter 1 (Sidebar 1-2). Pursuant to such plans, grants to states and communities from the NFMF are authorized to fund up to 75 percent of the cost of proposed flood mitigation activities.

The Midwest Floods: Mitigation in Practice

The Midwest floods of 1993 offered an opportunity to apply various types of post-disaster mitigation on a massive scale. The floods prompted presidential disaster declarations covering 525 counties in nine states. Total damage was estimated to fall between $12 billion and $16 billion, with federal costs amounting to about $4.2 billion (Platt, 1995, p. 26). Flood insurance payments only accounted for $293 million, or about 14 percent of total federal costs (of which about half was for basement flooding outside floodplains). As the floodwater receded in the fall of 1993, numerous studies and conferences were conducted to assess what went wrong. Foremost among these was the report of the Interagency Floodplain Management Review Committee (IFMRC, 1994; also known as "Galloway Report" after the committee's chair, Brigadier General Gerald E. Galloway, Jr.).

This report documented that, unlike coastal disasters, the 1993 flooding in the upper Midwest disproportionately affected low-income households and individuals. A comparison of census data for the flooded areas with data for nearby upland tracts disclosed that the flood victims were on the average, older, poorer, and more likely to live in mobile homes. Homes in the flooded area frequently had "market values of less than $25,000 and often as low as $10,000 or $5,000" (IFMRC, 1994, p. 7). Many of the victims were, in a sense, trapped; they lived in the floodplain out of economic necessity rather than by choice.

Before final publication of the Galloway Report, Congress in December 1993 acted to lessen future vulnerability by acquiring chronically flood-prone properties and helping their occupants relocate to safer areas (P.L. 103-181). Altogether FEMA bought out about 8,000 properties in nine states, at a total cost of more than $150 million.

Thousands of miles of nonfederal levees were breached or overtopped despite valiant efforts to defend them with sandbagging in many locations. However, as mentioned earlier, the Corps of Engineers reported that "all federally funded flood storage reservoirs operated as planned during the 1993 flood" and prevented $19.1 billion in losses (IFMRC, 1994, pp. 48 and 21). Corps reservoirs in the Missouri Basin stored about 18.7 million acre feet, reducing downstream flood damage on the Missouri by an estimated $7.4 billion. Corps levees on the Missouri River were estimated to have prevented another $4.1 billion in damages, primarily at Kansas City and St. Louis, where floodwalls were almost overtopped (IFMRC, 1994, p. 21).

Environmentalists and others maintained that future flood levels would be lower if nonfederal levees were either eliminated or allowed to fail in a large flood, and if substantial areas of floodplain drained for agriculture were allowed to revert to natural wetlands. The Galloway Report took issue with the argument that "flooding would have been reduced had more wetlands been available for rainfall and runoff storage" (p. 46). Modeling of the effects of wetland storage in four representative small watersheds indicated that "maximum reduction for floodplain wetlands was 6 percent of the peak discharge for the 1-year event and 3 percent of 25- and 100-year storm event[s]" (p. 47). Based on these findings, the committee concluded that "upland wetlands restoration can be effective for smaller floods but diminishes in value as storage capacity is exceeded in larger floods such as the Flood of 1993. Present evaluations of the effect that wetland restoration would have on peak flows for large floods on main rivers and tributaries are inconclusive" (p. 47).

On the subject of land use planning and regulation, the Galloway Report was politically cautious: "land use control . . . is the sole responsibility of state, tribal, and local entities . . . . The federal responsibility rests with providing leadership, technical information, data, and advice to assist the states" (p. 74). Despite this reluctance to endorse land use regulation in flood hazard areas, the report provided a balanced review of both structural and nonstructural issues pertaining to the recovery process.

The Oakland Fire: Rebuilding for Disaster

Despite all the emphasis on mitigation of multiple hazards in recent years, political, social, and economic forces conspire to promote rebuilding patterns that set the stage for future catastrophe. Perhaps nowhere has this been more conspicuous than the rebuilding in the Oakland–Berkeley Hills area after the firestorm of October 20, 1991, which destroyed over 3,300 homes in a few hours and killed 25 people. Few disasters have been more predicted beforehand or more analyzed afterwards. Yet the pattern of rebuilding, while safer in certain details, is more dense and congested than before and raises the prospect of another catastrophe. The next Bay Area earthquake is expected to occur on the

The Oakland–Berkeley Hills were rebuilt after the Oakland firestorm. Despite the very small size of lots in the burned area, reconstruction of homes within the same "footprint" was permitted. A new overlay zone for the burned area, adopted by Oakland 14 months after the fire, permitted enlargement of burned structures by 10 percent and exempted any plans submitted before its date of effectiveness, regardless of the size of structures proposed. The physical results of the ordinance are visible on the hillsides of Oakland: hundreds of very large, free-standing homes of eclectic design, often standing within 10 to 15 feet of each other on tiny lots, interlaced with roadways still as narrow and twisting as before the fire (Rutherford Platt).

Hayward Fault at the foot of the East Bay hills, "probably the most built-on fault in the world" (BAREPP, no date). About 1.2 million people live within the epicentral region of a potential 7.0 tremor on the Hayward Fault, ten times the population of 130,000 within the area affected by Loma Prieta (USGS Working Group, 1990, p. 4). The Hayward Fault has not seriously adjusted since 1868 and is estimated to have a 30 percent probability of a major earthquake in the next 30 years. Ground shaking in the East Bay would be at least 12 times worse in a 7.0 magnitude Hayward Fault earthquake than occurred there due to Loma Prieta (U.S. Geological Survey, 1994).

The Oakland fire of October 19–20, 1991, erupted under textbook conditions. Daytime temperatures hovered around 90°F and relative humidity was 17 percent. Hillside vegetation was bone dry. Dead plant material littered the ground and trees overhung many homes despite the long drought and warnings of fire danger. Hot, dry Santa Ana winds blew from the east on the morning of the conflagration. The California Department of Forestry had issued a "red flag" warning of potential fire hazard, but few residents took notice.

The conflagration began with a small brush fire. About 790 houses burned in the first hour after it began to spread. Turbulent winds generated by the fire itself spewed burning material in all directions. The fire crossed an eight-lane highway (Route 24) and continued to consume homes and vegetation further downslope. Public orders to evacuate were difficult to communicate in the absence of sirens. As the fire swept downslope, many tried to flee in their cars only to find it impossible to drive down the obstructed roads. Cars by the hundreds were abandoned as individuals literally ran for their lives, leaving the roads impassable to firefighters, who also soon lost all water pressure. There was little any of the victims could do at the last moment other than save themselves and whatever they could carry. Sixty years of building on the hills had created a hazard that no individual could undo. It was a classic "tragedy of the commons." All were swept up in the common peril, and personal consequences depended upon the fluke of the winds, not individual actions.

After the fire, all public officials proclaimed that the hills would be rebuilt. The option of not rebuilding but adding the burned area to the existing system of regional parks in the hills was not seriously considered. Public acquisition costs would have been sizable (estimated by the writer at about $400 million), and Oakland would have lost a valuable part of its tax base. Nevertheless, the failure to assess the economic and environmental impacts of not rebuilding represents a major deficiency in the post-disaster recovery process. Several studies by federal, state, and regional authorities provided much advice on decreasing fire risk in the area through use of tile roofs, improved firefighting capabilities, and management of vegetation. Many of these proposals were carried out at least partially.

But the major influence on the rebuilding process was the private market, fueled by insurance payments. One year after the disaster, 3,954 claims amounting to $1.4 billion had been filed with 49 insurance companies, averaging more than $350,000 per household. Reconstruction stimulated the largest building boom in California in 1992 and 1993, creating some 11,000 construction jobs during a time of statewide recession. Despite the very small size of lots in the burned area, reconstruction of homes within the same "footprint" was permitted. A new overlay zone for the burned area, adopted by Oakland 14 months after the fire, permitted enlargement of burned structures by 10 percent and exempted any plans submitted before its date of effectiveness, regardless of the size of structures proposed. The physical results of the ordinance are visible on the hillsides of Oakland: hundreds of very large, free-standing homes of eclectic design, often standing within 10 to 15 feet of each other on tiny lots, interlaced with roadways still as narrow and twisting as before the fire.

Hurricane Andrew: The Private Insurance Industry and Mitigation

The year after the Oakland fire, and three years after Hurricane Hugo and the Loma Prieta earthquake, the insurance industry was hit with another immense loss, this time in Florida. On August 24, 1992, Hurricane Andrew struck Florida south of Miami, and two days later it made a second landfall in Louisiana. Andrew was not strictly speaking either a coastal disaster or a flood disaster: the destructive force here was wind. Some 75,000 homes and 8,000 businesses in South Florida were destroyed or seriously damaged by sustained winds of more than 120 miles per hour. A total of over 250,000 people were displaced from their homes in Florida and Louisiana. Total economic loss due to Andrew was estimated at $30 billion. Total federal disaster assistance was about $2 billion, and the private insurance industry paid about $15.5 billion on 680,000 claims. Nine property casualty insurance companies were rendered insolvent by the disaster [Insurance Institute for Property Loss

Reduction (IIPLR), 1995a]. This was at the time the largest loss from a natural disaster in U.S. history (U.S. Congress, General Accounting Office, 1993, p. 4). (The 1995 Northridge earthquake was larger.)

Contrary to experience under the National Flood Insurance Program, Hurricane Andrew caused more damage to recently built structures than to those built before 1980 (IIPLR, 1995b). This partly reflected the use of building and architectural practices that actually increased vulnerability to wind damage. And about one-fourth ($4 billion) of the insured loss was attributed to shoddy workmanship and poor enforcement of building codes (IIPLR, 1995b).

Because of Hurricane Andrew, the private casualty insurance industry in the United States assumed a more active role in promoting safe building and land use practices. In remarks to the National Mitigation Conference in December 1995, Eugene Lecomte, CEO of IIPLR, stated, "Our mission is to reduce deaths, injuries, and property damage resulting from natural hazards—from hurricanes, earthquakes, tornadoes, floods, windstorms, hail, freezing, and urban wildfires. . . . If we can prevent building in the most hazard-prone areas, we can prevent property damage and probably eliminate many deaths and injuries. So why do we allow building on earthquake faults? On coastal shorelines where hurricanes hit? . . . All we ask is that builders in these areas acknowledge that their structures—whether homes or commercial buildings—are at greater risk and mitigate accordingly. *In some cases, that may mean not building at all in the most vulnerable areas.*" The insurance industry's reaction may signify a new era of hazard mitigation in the United States in which the industry presses the government to do what the property rights movement has argued against: regulate private land use to reduce disaster losses.

CONCLUSIONS

Since the seventeenth century, Europe and the United States have gained considerable experience in preparing for and mitigating the effects of major natural disasters. Elements of this preparation have included: (1) recognizing a disaster as an opportunity to alter laws, policies, and practices relating to the reconstruction process; (2) considering a range of approaches to reduce vulnerability, including both structural and nonstructural measures; and (3) recognizing the limitations of technology, especially with respect to lifelines.

Disasters inevitably raise the issue of the level of responsibility in regard to intervention to achieve mitigation—namely, *who should be required to do what*? The private and local levels of authority usually seek to at least rebuild the *status quo ante*, and preferably bigger and better. While well-intentioned efforts may seek to incorporate mitigation into the rebuilding process, the combined influence of private real estate values and municipal tax base concerns often stymie any significant changes in building patterns, as in the case of Oakland after its 1991 fire. (Even London in its 1667 Rebuilding Act prescribed more drastic replanning of the street plan than occurred in Oakland.)

Under the 1988 Stafford Act and its predecessors, the Federal Emergency Management Agency is the prime federal response agency. Mitigation has been a key element of national disaster policy and programs, as reflected most recently in the creation of a FEMA Mitigation Directorate in 1993. Yet FEMA today seeks to devolve responsibility for mitigation to lower levels of government under the rubric "all mitigation is local." Federal technical assistance and, in certain cases, funding are provided, as after the Midwest floods. But the initiative is left to nonfederal authorities to pursue mitigation efforts, even though they may conflict with the local and private interest in rebuilding as quickly as possible. As noted in the Galloway Report (p. 180), the assurance of federal assistance in the event of a repeat disaster creates a "moral hazard" by lowering the incentive to avoid risk.

In the case of flood disasters, the National Flood Insurance Program specifies more detailed federal standards for local mitigation than is the case under the Stafford Act. Yet as discussed earlier, land use planning and management have been subordinated to building and elevation requirements. Even in highly hazard-prone floodways and coastal areas, building and rebuilding are generally allowed under NFIP standards, and flood insurance coverage is available to the mean high-water line along tidal coasts, with the cost contingent upon elevation. This in turn obligates the private insurance industry to provide wind and comprehensive coverage to homes in hazardous locations. According to the Insurance Institute for Property Loss Reduction (1995b, p. 2), total privately insured property exposure values (not including the NFIP) in coastal counties increased by 69 percent between 1988 and 1993, from $1.86 trillion to $3.15 trillion (although much of this coverage applies to property not situated directly on the coast).

Certain states impose stricter requirements than the NFIP, such as Florida and North Carolina in regard to coastal erosion zones, and Illinois regarding 100-year floodplains. But the growing influence of the property rights movement, discussed in this chapter, may undermine the viability of such prudent limitations.

Thus, the issue of the appropriate scale of intervention seems to be an "Alphonse and Gaston" situation where each level of authority looks to the others to take leadership, provide funding, and accept the legal and political heat for prescribing unpopular or costly mitigation initiatives.

CHAPTER THREE

Governing Land Use in Hazardous Areas with a Patchwork System

PETER J. MAY AND ROBERT E. DEYLE

AT THE HEART OF THE THORNY problems posed by management of land subject to natural hazards lies a central conflict in public policy goals. On the one hand there is the goal of promoting economically beneficial uses of land, and the accompanying desire to allow individuals free use of their property, unimpeded by governmental intervention where possible. But these goals often conflict with the goal of promoting public safety and the welfare of the larger community through land use management policies that protect against the destructive effects of floods, coastal storms, earthquakes, landslides, and wildfires.

There are strong incentives to permit and even promote economically beneficial use of land despite the presence of natural hazards. The land may have significant value as residential real estate, or as rich, riverine floodplains, or as important commercial access to resources or navigation. Moreover, there is a reluctance to constrain or prohibit land use when areas are already in private ownership or have already been developed in economically beneficial ways.

On the other hand, private owners of land vulnerable to natural hazards may put their land to use in a way that threatens public safety. Some argue that government is obliged to protect individuals from their unwise land use decisions, or to protect their neighbors from the effects of those decisions as, for example, when filling floodplain land in one area results in increased flood risk or destruction of valued wetland habitats downstream. Still others maintain that it is inequitable for governments to subsidize land use in hazardous areas through public financing of emergency management and disaster response programs, or through public maintenance of the infrastructure that serves those areas.

In addition to the choices posed by these conflicting goals, policymakers must also decide upon the appropriate roles of different levels of government in land use and development management. At the heart of this choice are issues of the ability and willingness of different governmental levels to undertake such management. State governments delegate most decisions about land use to local governments, although federal and state interests in reducing disaster losses and minimizing relief payments in the aftermath of disaster provide a compelling rationale for higher-level government involvement. Relationships among different levels of government involved in land use management are complex, with a number of governmental levels and agencies sharing responsibility in relation to various hazards. Figure 3-1 illustrates these relationships with regard to earthquake and flood hazards.

As discussed in Chapter 2, the federal government has assumed increasing responsibility for financing structural protection against natural hazards and supplementing state disaster-relief efforts. States also have assumed a sizable burden of the nonfederal share of the costs of disaster response and recovery. Generally, federal and state governments over the past two decades have become increasingly involved in directly regulating land use and in influencing land use decisions by local governments.

The various choices made by the federal and state governments in attempting to influence land use and development in hazardous areas are reflected in a diverse set of policies and programs. Most of these are aimed at influencing the ways in which local governments regulate land use or development. The various policies and programs resemble a house that has been subjected to numerous cosmetic fixes and redesigns over the years. Different owners and architects have added their perspectives on what would make it more livable. New additions have been appended to the old structure to accommodate changing demands or preferences and different ideas about what constitutes good design. As a

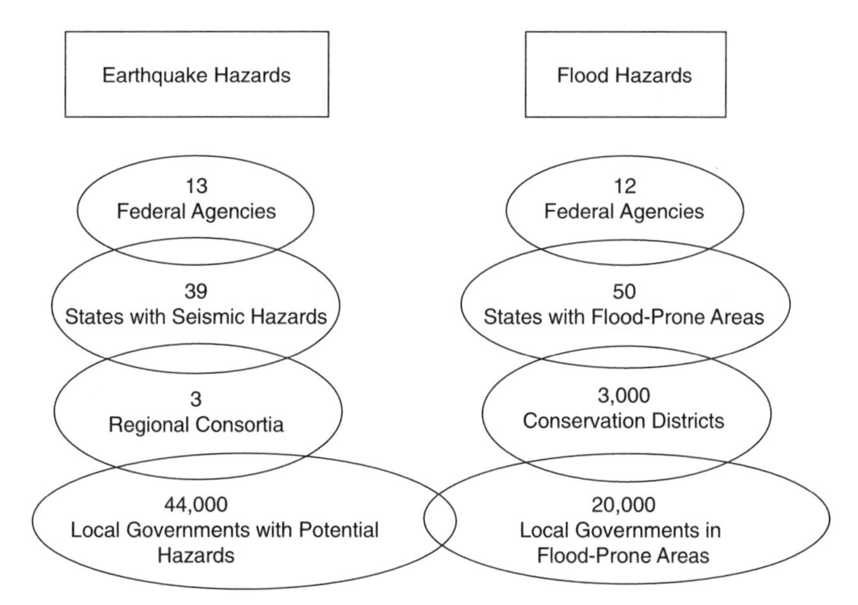

FIGURE 3-1 Relationships among different agencies and levels of government can be complex when they share responsibility regarding earthquake and flood hazards. Some government levels have jurisdiction over areas that are both earthquake- and flood-prone. Sources: Federal Interagency Floodplain Management Task Force (1992); Federal Emergency Management Agency (1994c).

consequence, the system is awkward and inconsistent in design, and many of its parts do not function well. The house is better thought of as a set of overlapping structures than the product of any grand design.

In recent years there has been increasing criticism of the patchwork of federal and state policies and programs governing local decisions about land use and development. Concerns about the costs imposed by the patchwork—both compliance costs and lost income caused by delayed or restricted development—have given way to a much stronger reaction expressed as a movement to protect private property rights. In addition to these criticisms, the hodgepodge of land use policies and programs is now being measured against a new standard—environmental sustainability. The critical test is whether the system helps the private sector and local governments make choices about development and growth that ensure long-term maintenance of a desired quality of life. Fostering sustainability fundamentally challenges intergovernmental policy design; it requires examination of the complexities of human interaction with the natural environment over the long term, and consideration of the cumulative effects of development on air, land, and water, and among different jurisdictions. Such an examination also raises issues about the appropriate spatial and temporal scales of planning and policy making.

This chapter discusses the choices faced by various levels of government and the resultant intergovernmental patchwork of policies and programs governing land use and development in hazardous areas. The primary focus is the intergovernmental system as it relates to decisions that the private sector and local governments make about land use and development. We also look at the motivations and interests of different levels of government, potential conflicts among and within these levels, and the implications for the design and implementation of natural hazards policies.

LOCAL GOVERNMENTAL CHOICES: RELUCTANCE TO ENACT STRONG LAND USE MEASURES

If people chose not to locate in hazard-prone areas, the problems posed by development would be greatly diminished. However, the reality is that certain types of hazardous areas are often among the most desirable for development. Moreover, development of these areas has been made possible by engineering advances that allow building on steep hillsides and in areas with poor soils; by construction that is more resistant to wind, flooding, and earthquakes; by protective works that limit risks posed by flooding and coastal storms; and by risk-sharing mechanisms such as federally sponsored flood insurance and federal disaster assistance programs that reduce the private risks of developing hazard-prone areas. Given these factors, landowners and developers are unlikely to alter their preferences for development of such land without prodding.

Such prodding might be expected from local governments, but it is not a simple matter to put in place a set of rules that balance desires for development with needs for protection of life, safety, property, and environmental quality. Achieving this balance often pits powerful forces against each other (see Logan and Molotch, 1987). Various studies point to the reluctance of local governments to adopt, or adequately enforce, strong measures for managing land use and development in hazardous areas. National studies of risk reduction measures by Berke and Beatley (1992b), Burby and French et al. (1985), Godschalk et al. (1989), and Wyner and Mann (1986) document noteworthy gaps in local policy adoption of land use measures for reducing risks posed by earthquakes, floods, and coastal hazards.

Local governments are often reluctant to adopt such policies for a variety of reasons. First, like individuals, unless governments have direct experience with the devastation wrought by a natural disaster, they tend to discount the risks involved in allowing development in hazardous areas. Second, even in relatively high-risk areas where losses have already occurred, other problems are often higher on local agendas. Third, hazard-prone areas often are very valuable economic resources because of natural amenities such as open topography, view, ocean frontage, and access to water and water-based transportation. Finally, because hazard-prone areas often are already built out, remedial actions for addressing hazards are costly to implement and politically difficult. As a consequence, local governments are typically reluctant to take strong steps to regulate land use and prefer other hazard mitigation mechanisms such as structural protection or strengthening existing buildings. Federal programs reinforce this preference by allowing costs to be shifted from local taxpayers to the federal treasury.

Two factors are key to local governments' use of strong land use planning and development management programs for hazard mitigation. Of prime importance is the *commitment* of local officials to manage development in hazard-prone areas. The willingness of local elected officials to advocate such measures is an essential ingredient to local governmental action, and their reluctance to do so serves as a key impediment. Also relevant, but generally found less of a constraint than commitment, is the *capacity* of local governments. While lack of resources and technical expertise can impede such programs, shortfalls in capacity are generally not an insurmountable barrier in larger jurisdictions, with the notable exception of building code enforcement (see Federal Emergency Management Agency, 1992c, pp. 74–76; Berke and Beatley, 1992b, p. 94). The importance of commitment and capacity is documented in a number of studies, but stands out most clearly in work by planning scholar Raymond Burby and his colleagues on local planning for hazard mitigation (see Burby and May et al., 1997; Dalton and Burby, 1994; May et al., 1996). One clear implication of this line of research is that the willingness of local governments to undertake risk reduction programs has more to do with community resources and the extent of local political demands, which affect the commitment of elected officials, than with previous experience with disasters or objective risk.

Clearly there are constraints on the ability to reduce the risks of natural hazards through land use planning and development management. The existing patterns of development and the nature of the hazard limit the appropriateness of different planning tools. A community in which a large amount of land is susceptible to severe risks is much less able to use land use tools to steer development to safer land. Also, communities with large areas already developed will be less able to apply land use measures as a primary mitigation tool.

Studies of post-disaster recovery further demonstrate the reluctance of local governments to significantly restrict land use in hazardous areas even when the risks of such land use have been vividly demonstrated (Haas et al., 1977; Rubin et al., 1985; Rubin and Popkin, 1990; Smith and Deyle et al., 1996). After a disaster, local governments tend to focus on returning to normal as quickly as possible. Where there is an attempt at planning the post-disaster recovery process, the focus is typically on improving the community by enhancing its economic viability. Mitigation measures—either structural alterations to the environment (dams, levees, etc.) or strengthening of the built environment—are often employed to protect against future hazards. Radical changes in land use are rarely attempted and even less frequently accomplished. There have been few instances where a post-disaster recovery plan was in place prior to a natural disaster.

FEDERAL CHOICES GOVERNING STATE AND LOCAL ACTIONS

The federal government has a strong interest in reducing disaster outlays by promoting land use and development measures that reduce exposure to hazards. However, the federal government depends on local government to put hazard mitigation measures in place, and local governments have been reluctant to play this role (more generally, see May and Williams, 1986). The states may be able to resolve this impasse by acting as intermediaries in bringing about local action. In this section, we consider federal choices and influence.

Federal (and state) policymakers can make use of a variety of interventions to influence land use and development (see Table 3-1). Generally, these interventions fall into two categories: direct and indirect involvement. Higher levels of government can act directly to regulate land use and development, or they can attempt to influence development directly by investing in real property or infrastructure. Acting more indirectly, higher levels of government can influence decisions of lower-level

TABLE 3-1 Federal and State Choices for Influencing Land Use and Development

Direct Intervention	Indirect Intervention	Incentives and Information Overlays
Land use regulation	Planning mandates	Financial assistance
Investment in land or infrastructure	Regulatory mandates	Technical assistance Education and training Information

governments by imposing planning requirements or other mandates. Other policy tools include the use of various incentives (e.g., funding, technical assistance), information (e.g., maps of hazardous areas, risk profiles), and education and training.

The Federal Patchwork of Programs

A plethora of federal programs attempt to influence decisions by the private sector and local governments about land use in hazardous areas. Direct federal regulation of land use has been limited to the protection of wetlands and endangered species, and has had only limited impact on reducing the risks of natural hazards. For some hazards, the federal government has employed regulatory mandates and incentives to prod local governments to initiate controls over land use and development, or to at least analyze the hazards presented by development. For other hazards, the federal government has used investment and incentive strategies to influence private sector development behavior. In still other cases, federal influence has been limited to the provision of technical information or financial support for planning or development management by states or local governments.

With the exception of wetlands permitted under Section 404 of the Clean Water Act (P.L. 95-217; *U.S. Code*, vol. 33, sec. 404), the related "swampbuster" provisions of the 1985 Food Security Act (P.L. 99-198; *U.S. Code*, vol. 16, sec. 3821), and the restrictions on habitat modification under the Endangered Species Act (P.L. 93-295; *U.S. Code*, vol. 16, sec. 1534), the federal government has avoided direct regulation of land use. Wetlands regulations, in fact, are grounded in the federal government's authority to regulate navigation under the interstate commerce clause of the Constitution (Article 1, Section 8, Clause 3) rather than any explicit intent or authority to regulate land use (see Ferretti et al., 1995, p. 12-139). One could argue that restrictions on wetland development have reduced exposure to flood hazards where the flood storage capacities of wetlands have been preserved in riverine floodplains and in coastal areas at risk from hurricane-driven storm surges. However, such impacts were not the principal focus of the legislation and do not accrue from the protection of all types of wetlands.

Provisions under the National Flood Insurance Act of 1968 that require local governments to adopt building codes for floodproofing and elevation of habitable structures constitute a regulatory mandate. These provisions were strengthened with the enactment of the Flood Disaster Protection Act of 1973 which requires communities to adopt the specified regulatory measures in order to be eligible for federal aid, including disaster assistance for flood damages, loans for rebuilding after disasters, and eligibility for federal flood insurance. As noted in Chapter 1, these incentives served to induce much greater participation in the federal flood insurance program than was the case prior to their imposition. By 1978, over 85 percent of the local governments in the nation with designated flood hazard areas were participating (Petak and Atkisson, 1982). Although the program has had an important impact in stimulating local action, rather than emphasizing restriction of development (outside of the floodway itself), the emphasis has been on construction standards. Some observers (e.g., Burby and French, 1981; Platt, 1987) argue that this emphasis, coupled with the availability of relatively low-cost flood insurance, has encouraged floodplain development.

Since the late 1980s, the Federal Emergency Management Agency has offered incentives under the Community Rating System to encourage greater local effort to avoid development of floodplains. Local risk reduction initiatives that exceed the minimum requirements under the Flood Insurance Act—such as more extensive regulation of flood hazard areas, mapping, public information, and flood damage reduction—can result in lower flood insurance premiums for property owners. While participation and hazard reduction have been modest, community effort to mitigate flood hazards has been recognized with this program.

Executive Order 12699, issued January 5, 1990 (3CFR, 1990 comp., p. 269), is an example of federal leadership in hazard mitigation. In an attempt to call attention to seismic issues in new construction, the executive order requires that new buildings constructed and leased by the federal government comply with appropriate seismic design and construction standards. It also requires federal agencies that are responsible for financing new construction (i.e., through federal loan or mortgage insurance programs) to develop plans for using appropriate seismic measures in construction.

The Coastal Barrier Resources Act (CBRA) (P.L. 97-348; *U.S. Code*, vol. 16, sec. 3503), enacted in 1982 and revised in 1990, employs a mix of investment and incentive strategies for regulation of development on coastal barrier islands. The act designates coastal areas as ineligible for federal subsidies for growth-inducing infrastructure such as roads, bridges, water, sewer systems, and protective works. In addition, new development within such areas is ineligible for federal flood insurance. While CBRA has accomplished the objective of reducing federal financial exposure to the risks posed by coastal hazards, the act has had limited success in actually deterring development (Godschalk, 1987; U.S. Congress, General Accounting Office, 1992). Development has continued in some coastal barrier islands with private financing, especially of high-valued projects such as multi-story condominiums.

Other federal programs operate indirectly to influence state and local policy on land use and development in hazardous areas. The Coastal Zone Management Act of 1972 (P.L. 92-583; *U.S. Code*, vol. 16, sec. 1451 et seq.) declared a national policy favoring better management of coastal land and water resources and authorized coastal states to develop and implement coastal zone management plans (Godschalk, 1992). To encourage states to prepare management plans for the coastal zones, the act offered incentives including federal funding for the design and implementation of federally approved plans. It also granted states the power to review federal actions occurring within the designated state coastal zones. State plans were required to address nine areas of declared national interest, two of which concern natural resource protection and hazard management. As noted by Godschalk (1992), the $696 million invested in the coastal program over its first 14 years had a substantial impact, motivating 29 states to design and carry out coastal management efforts. But because several basic aspects of the coastal plans are subject to state determination (e.g., the definition of the coastal boundaries, permissible uses, intergovernmental roles), there is considerable variation in approach from state to state. Moreover, the voluntary nature of the program leaves room for considerable variation in overall state effort.

Planning is also required under provisions of the Stafford Act of 1988, which amended the 1974 Disaster Relief Act. States are required to prepare and implement "Section 409" hazard mitigation plans as a condition for receiving disaster relief grants or loans. The plans must include an evaluation of the options for mitigating the natural hazards that threaten designated areas in the state, and must be revised following disasters that qualify for presidential declarations. The Federal Emergency Management Agency considers the status and implementation of these plans when evaluating a state's application for disaster assistance. However, local government participation in this process is at the behest of the responsible state; there are no explicit incentives for local governments to consider the provisions of the state plans in their own comprehensive planning. Reviews of these state plans have generally been critical of their failure to tie the nature of the problem to stated goals and objectives, to identify priorities in hazard mitigation, and to adequately relate priorities to implementation strategies (Deyle and Smith, 1994; Godschalk, 1995).

Other federal programs employ a mix of incentives, information, and education to influence local decisions about land use and development in hazardous areas. With regard to earthquakes, the Federal Emergency Management Agency, in conjunction with the Building Seismic Safety Council, a nongovernmental organization, has encouraged the development of earthquake-resistant construction standards and seismic standards for lifelines. FEMA has provided technical and financial assistance for federal-state partnerships for fostering stronger earthquake-preparedness efforts. The U.S. Geological Survey has attempted to mobilize local planning and response efforts by documenting seismic risks in urban areas and developing tools that local governments can use in their land use planning.

In sum, a wide range of federal programs addresses land use and development in hazardous areas. The federal government plays a direct regulatory role in wetlands protection. Through the Coastal Barrier Resources Act the federal government uses investment policies to deter development of hazardous coastal areas. With regard to earthquake and flood protection, the federal government does not intervene directly but attempts to induce state and local governments to carry out national objectives. The National Flood Insurance Act, for example, offers the incentive of affordable flood insurance as an inducement for communities to regulate development in floodplains. The Stafford Act of 1988 was enacted to bring about a more comprehensive and coherent approach to achieving mitigation of a number of natural hazards. Yet, the resulting effort—like previous efforts to provide integrated approaches—has encountered obstacles such as lack of standards for planning, cumbersome bureaucracy, and the indifference of states and localities, or their inability, to engage in the desired planning processes.

The Legacy of the Federal Patchwork

Several key points can be made about the diverse federal programs and policies that relate to the management of natural hazards. First is the patchwork nature of federal programs and policies. No overarching federal policy governs land use and development in hazard-prone areas. Instead there are individual initiatives such as the Coastal Zone Management Act addressing development in coastal areas; the Coastal Barrier Resources Act addressing development on designated barrier islands; the National Flood Insurance Act addressing development in flood-prone areas; Section 404 of the Clean Water Act and the "swampbuster" provisions of the 1985 Food Security Act addressing development in wetlands; various disaster-relief provisions of the Stafford Act that open opportunities for mitigation in the aftermath of major disasters; and several technical and financial assistance initiatives concerning seismic hazards. In total, there are over 50 federal laws and executive orders that relate to hazard management (Federal Interagency Floodplain Management Task Force, 1992).

Secondly, land use provisions are often the weak elements of federal programs; they are the provisions that are given low priority, are ignored and unfunded. The Coastal Zone Management Act allocates funds to states for developing plans for managing growth in coastal areas, but for most states participating in the program, planning for natural hazards has taken a back seat to other goals such as accessibility and environmental quality. The Water Resources Development Act of 1974 requires the Army Corps of Engineers to consider nonstructural alternatives to prevent or reduce flood damages (P.L. 93-251), but this provision largely has been ignored (Burby and French et al., 1985). The National Flood Insurance Act included land use planning provisions, but these were never implemented.

A third key point is that the net effect of federal programs is to encourage development in hazardous areas. For example, rather than prohibiting or restricting development in flood-prone areas (with the exception of floodways), the National Flood Insurance Program has arguably encouraged development, although development is elevated to protect against losses (i.e., from a 100-year storm). By encouraging development, the NFIP has increased the risk of losses from more severe,

Sandbags and pumps keep floodwater away from residential area, St. Genevieve, Missouri (FEMA).

less frequent storms, such as the Midwest floods of 1993. This bias toward solutions that encourage development and away from land use approaches also characterizes the current federal involvement in the reduction of earthquake hazards. As noted above, federal initiatives aimed at mitigating earthquake losses consist of a mix of developing new building code standards, providing information and education about earthquake hazards, and financing and participating in collaborative-planning partnerships. The potential for land use provisions has been overshadowed by efforts to foster stronger building codes, to encourage enactment of such codes by states and localities that do not have them, and to improve code compliance and enforcement.

A fourth key point about the federal patchwork is that a host of other federal policies can limit opportunities for local governments to use land use management tools for hazard mitigation. The largest influence by far has been the extensive construction of protective works along major rivers, first mandated by the Flood Control Act of 1917 (*U.S. Statutes at Large*, vol. 39, p. 948), then in the 1930s by a series of amendments to the Rivers and Harbors Act of 1899 (*U.S. Statutes at Large*, vol. 30, p. 1121). Despite the expenditure of many billions of dollars for structural flood control measures, flooding remains a formidable hazard. Moreover, like the provisions of the National Flood Insurance Act, the protective works have arguably increased the numbers of people and the amount of public and private property exposed to catastrophic flooding. The Federal Interagency Floodplain Management Task Force summed up the situation as follows: "Floodplain management should result in an actual decline in the nation's flood losses, including public and private property damage, injuries, and disaster relief. This has not been achieved. In fact, there was a definite increase in flood damages from 1916 to 1985, although there is evidence that these losses have remained fairly constant over the last two decades when compared to broad economic indicators like the GNP" (1992, p. 60). Increased exposure of people and property to flooding has been abetted by federal financing of highway construction, sewers, and other infrastructure that increase the likelihood of development in flood-prone areas while reducing development costs.

To sum up, the federal effort to date has for the most part failed to advance land use planning to manage losses from natural disasters, and to promote sustainability. The patchwork of federal programs affecting land use in hazard-prone areas has resulted in missed opportunities, notably, the failure to take advantage of land use provisions in the National Flood Insurance Program, or to make stronger land use provisions part of federal policy regarding other hazards. Moreover, a number of long-standing federal programs contain a bias against effective use of land use methods for hazard mitigation; these programs show a strong preference for protective works which in the long run increase rather than decrease the potential for catastrophic loss.

STATE CHOICES: GOING THEIR OWN WAY

With the exception of regulatory actions spurred by the National Flood Insurance Act and the planning incentives of the Coastal Zone Management Act, states have largely gone their own way in devising policies governing land use and development in hazardous areas. Not surprisingly, there has been a wide divergence in the approaches taken by states and in the extent to which they actively seek to mitigate risks. Like the federal government, states have a number of choices concerning when and how to intervene in decisions about land use and development. States can intervene directly and exert their regulatory powers over local decision making. Or states can establish planning or other processes that require or encourage local governments to take actions in concert with state desires. Or states can choose to ignore problems posed by development in hazardous areas and leave them for other levels of government to address. As is true at the federal level, a variety of policy instruments can be incorporated into state policy.

The Range of State Programs

The choices that states have made span the full range of policy approaches. Some states, such as Delaware, Florida, Michigan, New York, North Carolina, and Wisconsin, have been willing to intervene directly in private land use decisions by imposing requirements upon development of environmentally sensitive or hazardous areas such as wetlands and coastal zones. Other states have required local governments to enforce provisions of state building codes that address seismic hazards or hurricane winds. In addition, Florida, North Carolina, Washington, and others have established planning mandates, with varying levels of incentives, to prod local governments into considering natural hazards as part of comprehensive planning. A few states, such as California, Florida, Massachusetts, and North Carolina, have used their investment policies for land acquisition and public infrastructure to influence land use and development in hazardous areas.

By far the most common approach chosen by states is direct regulation of selected areas (e.g., wetlands) or aspects of development (e.g., coastal development). This entails state promulgation of rules governing land use or development for the particular situation, and state responsibility for enforcing the rules (which may be delegated to local governments). For example, Godschalk et al. (1989) report that all 18 hurricane-prone coastal states have enacted some type of direct state regulation regarding beach erosion, wetland areas, or sand dunes. At least 13 states now impose various coastal setbacks, requiring new development to locate a certain distance inland from the shoreline. Similarly, California has a requirement that no structures for human occupancy can be placed within 50 feet of a defined earthquake fault. In addition, many states have established special regulations governing development near sensitive areas such as wetlands (see Kusler, 1980).

More than a dozen states have enacted mandates that require local governments to develop comprehensive land use plans. The states specify policy goals and objectives but, to varying degrees, leave the specific details of the content and implementation of plans to local governments.

The state programs vary in the extent to which they identify problems posed by development in hazardous areas to be considered in the local planning process. As noted by Burby and May et al. (1997), there are three basic arguments for the comprehensive-planning approach. First, in some states local governments have shown a willingness to grapple with land use problems and are prepared to work as partners with state governments in addressing these problems. Second, the emphasis on comprehensive planning recognizes the interrelated character of land use decision making and other aspects of community development. A third argument is that comprehensive-planning programs can overcome shortcomings that sometimes characterize direct state regulation of land use or development. These shortcomings or obstacles include inefficiencies inherent in uniform state standards, objections to state control over local decisions, gaps in piecemeal approaches, and difficulties in monitoring and enforcing state standards.

California has mandated local planning since 1937, but it was not until 1972 that California required local governments to consider natural hazards in their planning. Subsequent development of state planning mandates has occurred in two waves. In the 1960s and early 1970s the California mandate was revised, and new programs were established in

These houses on pilings survived a nor'easter in Scituate, Massachusetts, 1991 (Ralph Crossen, Building Commissioner, Barnstable, Mass.).

Florida, Hawaii, Oregon, and Vermont. Also at this time, planning requirements for selected substate areas were mandated in Colorado (unincorporated areas of counties), Massachusetts (Cape Cod Commission, Martha's Vineyard), New Jersey (Pinelands area), New York (Adirondack Park), and North Carolina (coastal counties). The second wave of planning mandates began in the mid-1980s and included major revisions to planning requirements in Florida and Vermont, and the institution of new state mandates in Delaware, Georgia, Maine, Maryland, New Jersey, Rhode Island, and Washington.

States have also employed a mix of incentives and informational and educational approaches to influence local decisions about land use and development in hazardous areas. One approach is to withhold funds for public facilities or other infrastructure in such areas so as to discourage development. Florida's policymakers have enacted strong legislation of this type, which substantially restricts public investment in hurricane-prone coastal areas. Other states that use variants of this general approach include Delaware, North Carolina, Massachusetts, and South Carolina (see Godschalk et al., 1989, and Deyle and Smith, 1994). States also offer incentives to local governments in the form of low-interest loans and grants for funding mitigation planning and programs. This is particularly common in the funding of flood control projects; for example, the Texas Water Control Revolving Fund is a source of funding for development of structural and nonstructural protection against floods. Florida recently initiated a matching grants program to support preparation of local hazard mitigation strategies. Many states provide information, such as maps or technical assistance, in order to enable local governments to carry out state objectives. In California, for example, the Division of Mines and Geology has produced maps of ground shaking, landslide, and liquefaction potential for some localities, and the Division of Forestry has produced wildfire hazard maps for urban areas of the state.

In sum, states, independently from federal initiatives, have developed an array of policies and programs relating to hazardous areas. There is considerable variation in the specifics of the state regulations, mandates, and other strategies that have been adopted. While most states have policies or programs that influence development within environmentally sensitive areas like wetlands, fewer states have taken initiatives that are explicitly concerned with development in areas subject to natural hazards. However patchy, there clearly has been more progress at the state level than at the federal level both in using land use measures to manage development in hazard-prone areas and in stimulating local efforts to better manage development in such areas.

The Legacy of Divergent State Choices

Perhaps the most noteworthy point about state management of natural hazards is the diversity of choices that have been made. In studying state mandates governing hazard-prone areas, May (1994) found several reasons for state choices. State legislatures vary in their willingness to intervene in local decision making, reflecting differences in the political culture of the states. Willingness to intervene at the local level is, in turn, affected by the level of agreement among state policymakers about the seriousness of the hazards. Where agreement is strong, there is a predisposition to craft a tough statute. Another source of variation in state choices arises from within-state differences in the political power of various development and environmental interests.

Although a number of states have undertaken direct regulation of various aspects of development, there is a notable lack of systematic evaluation of those experiences. In one notable exception, a study of programs in 20 states to reduce erosion and sedimentation pollution in urban areas, Burby (1995) found a number of deficiencies in program content and performance. Chief among these, as the program administrators asserted, were low levels of staffing and funding relative to the tasks at hand. Recognizing the limited capacity of state agencies, Burby and Paterson (1993) found that local enforcement of a North Carolina program at least equaled, if not surpassed, state performance.

Planners have taken more interest in state mandates requiring local governments to undertake comprehensive planning. Dozens of articles and several books have described these mandates and the state land use management systems in which they are embedded, and have compared programs across states (for an overview, see Bollens, 1993). But, except for critical commentary about efforts of states in the 1970s to put in place comprehensive planning mandates, there has, until recently, been little assessment of the impact that these state mandates have had on local government policy and land use.

A national study of state comprehensive planning mandates (Burby and May et al., 1997) reached more positive conclusions than evaluations of the first generation of state planning mandates. By comparing the actions of local governments in states that have imposed planning mandates with the actions of local governments in states that have not imposed such mandates, Burby and his colleagues show that the planning mandates have stimulated local government efforts to plan for and manage development in hazard-prone areas. Local governments are more likely to prepare comprehensive plans when required to do so, and state planning mandates foster a substantial improvement in the quality of plans and their attention to natural hazards. For states that require comprehensive plans of local governments and follow through on those requirements, local plans are more well grounded in fact, they state goals more clearly, and they propose stronger local policies for guiding development.

However, while it is better to have a state comprehensive planning mandate than not, such mandates do not guarantee local attention to hazards. In a detailed study of the degree to which local plans comply with Florida's planning mandate, Deyle and Smith (1994) found substantial variation. Policy choices made by the state implementing agency had a significant impact on which hazard mitigation planning mandates were complied with by local governments. This study also documents substantial tolerance by the state for variation among communities in their compliance with most of the hazard mitigation planning requirements.

Similar variation in local government compliance with state mandates is documented by May and Birkland (1994) in a study of earthquake policies in California and Washington State. A higher percentage of local governments in California adopted earthquake risk reduction measures than in Washington State. California has both a number of state mandates governing such actions and a history of notable earthquakes. Washington State has similar seismic potential, but less extensive earthquake experience and fewer mandates. The key finding from the study, however, is that compliance with California's earthquake-specific mandates is uneven; much slippage occurs at the stage of actually carrying out the policies. In their analysis of the data from the comparative state survey conducted by Burby and his associates, Berke et al. (1996) document substantial variation among the local plans prepared both with and without state planning mandates.

The research reviewed here on state planning and hazard reduction mandates is sobering in that it shows that such mandates, in stimulating local efforts, produce marginal rather than widespread changes. Bringing about major shifts in land use or development decision making is very difficult given existing development patterns and the entrenchment of forces for development. The influence of planning mandates varies considerably among the states due to differences in policy design and the strength of efforts by relevant state agencies to implement mandates. By far the most important factors in enhancing local attention to these issues are strong commitment by relevant state agencies to the mandate, backed by strong implementation efforts. Florida's success in mandating comprehensive growth-management planning by local governments illustrates the importance of commitment-building incentives within mandates (e.g., threatening local governments with sanctions for failing to meet deadlines, requiring revisions of unacceptable local plans), accompanied by strong state implementation efforts.

The bottom line is that state governments have made progress in advancing local land use planning for natural hazards, but with notable gaps. The evidence is that state comprehensive-planning mandates can be useful in influencing the way local governments address development in hazard-prone areas. In particular, planning mandates draw attention away from limiting harm—for example, through hazard protection structures—toward preventing harm through management of land use. But, less than one-third of the states have chosen this route, and among those states there is substantial variation in outcomes.

CHOICES ABOUT REGIONS: THE LIMITATIONS OF REGIONAL LAND USE SOLUTIONS

One of the more challenging aspects of managing hazards is choosing an appropriate spatial scale. A strong case can be made for regional approaches, which have been advocated for the management of ecosystems, earthquake hazards, and flood hazards. Regions whose jurisdictional boundaries coincide with natural boundaries are ideal for managing natural resources and environmental conditions. This is the logic of watershed associations and river basin authorities that have been created for water resource planning and development (Platt, 1987). This kind of spatial scale is also illustrated by regional park management districts (such as the regional authorities created for managing natural resources in the Adirondack Mountains in New York State and the Pinelands in New Jersey) and regional districts for air pollution management and flood control in a number of metropolitan areas.

The United States is perhaps exceptional in the variety of forms of regional entities and special districts that have been created for dealing with such issues. The success of regional initiatives for land use planning has been limited however. A key impediment to the formation of regional levels of authority, whether in the form of entirely new organizations or less formal councils or federations, is that they pose a threat to other levels of government and their constituencies. No matter how desirable it is to have regional management entities, they constitute another layer of bureaucratic intervention and their existence can provoke turf wars. Local officials see funds going to regional organizations that could be coming to them, and they are concerned with potential encroachment on local authority. Rather than easing intergovernmental relations, regional initiatives can increase the potential for conflict and for fragmenting program implementation.

Florida's experience in carving out a broad-purpose regional role in land use planning illustrates the tensions that attend regional planning and management initiatives. There was long-standing tension between local governments and the independently authorized and funded regional water-management districts. Partly to avoid exacerbating these conflicts, Florida's policymakers assigned regional planning functions under the state's 1985 growth management legislation to a different set of regional authorities, involving regional planning councils that had less power. Even though local officials hold two-thirds of the seats on each regional council (the other third are appointed by the governor), local governments resent regional interference in local affairs, just as they resent state interference. The precariousness of the regional planning function was revealed by efforts in the 1992 legislative session to abolish regional planning councils. This eventually resulted in legislation redefining their roles in order to minimize potential interference with local planning.

The experience of the federal government in creating regional entities has also been unsatisfactory. Efforts to mandate the creation of regional planning entities by lower levels of government, best exemplified by the Section 208 areawide water quality planning provisions of the 1972 Federal Water Pollution Control Act Amendments (P.L. 92-500, *U.S. Statutes at Large,* vol. 86, p. 816), have been largely unsuccessful (Deyle, 1995). The six regional river basin commissions created by the 1965 Water Resources Planning Act were never granted adequate resources to carry out their tasks (P.L. 89-80, *U.S. Statutes at Large,* vol. 79, p. 244). Furthermore, they were criticized for studies and plans that were too general in scope (Kusler, 1985). The Reagan administration dismantled these commissions in the early 1980s. In addition, Burby and French et al. (1985) found that regional councils did a poor job of floodplain management. After a survey of 585 regional councils, the authors concluded that most devoted relatively little of their overall staff resources to management of flood hazards. The majority of such agencies performed planning, coordination, and capacity-building functions. Very few exercised direct land use regulatory powers or undertook investment strategies such as land acquisition.

The most successful examples of regional planning and management of land have been new special-purpose organizations created by federal or state legislatures with the authority to directly implement or compel implementation of their plans and policies. Regional partnerships between levels of government that are less oriented toward intervention have also been successful in facilitating planning for natural hazards. Prominent examples of regional planning and regulatory entities include the Adirondack Park Agency, the New Jersey Pinelands Commission, the Tennessee Valley Authority, Florida's water management districts, the Delaware River Basin Commission, and flood control agencies such as the Denver Urban Drainage and Flood Control District. Platt (1986) attributes the success of the regional flood control agencies to a mix of fiscal autonomy, legal flexibility in interpreting their mandates, professionalism among staff, and clear goals for the agencies. In many instances, however, the political, organizational, and environmental conditions that facilitated such initiatives were exceptional.

Less ambitious regional partnerships between federal and state governments have been successful in enhancing the capacity of lower levels of government to plan for and manage natural hazards. Examples include the Southern California Earthquake Preparedness Program and the San Francisco Bay Area Regional Earthquake Preparedness Project, which were established by the Federal Emergency Management Agency and the state of California in the 1980s to provide technical information and technical assistance to local governments. The success of these programs has led to similar initiatives elsewhere around the country (Berke and Beatley, 1992b).

Some observers have argued that fragmentation among levels of government within a federal system is a fact of life that cannot be changed. Thus, as noted by Deyle (1995), collaborative alliances among existing government entities—such as the successful Chesapeake Bay Program and the regional earthquake partnerships between the Federal Emergency Management Agency and states—are probably the best that can be achieved in efforts to integrate natural-resource and land use management on a regional scale. The function of regional entities in governing land use in areas prone to natural hazards is likely to remain limited to broad-brush planning, intergovernmental coordination, and capacity-building functions of providing information, education, and technical assistance to local governments.

PROMISING FUTURE DIRECTIONS

The historic reluctance of federal and state governments to intervene directly in land use decisions by private property owners and the current movement toward reasserting property rights contribute to a political environment in the latter half of the 1990s that is not particularly supportive of a stronger federal or state regulatory presence in governing land use or development in hazardous areas. Given this environment, and the limited impact of previous federal, state, and regional initiatives that have relied principally or solely on incentives, investment policies, or information and education, the challenge is to identify ways to enhance the commitment of local officials to managing development without engendering such negative reactions. Many argue that future directions for environmental policy in general should include less emphasis on regulatory prescription and greater reliance on local governments as partners with state and federal governments in pursuing paths to sustainability—in other words, a more flexible form of intergovernmental policy mandate.

Various authors have written about collaborative forms of environmental management under the labels "co-production" (Godschalk, 1992), "collaborative planning" (Bollens, 1993), "civic environmentalism" (John, 1994), and "cooperative intergovernmental policies" (May et al., 1996). Although the use of these terms varies considerably, the central thrust is to devise higher-level policies that enhance local government interest in and ability to work toward achieving policy goals. These programs may prescribe planning or process elements to be followed (a form of policy mandate), but they do not prescribe the particular means for achieving desired outcomes. Cooperative mandates use financial and technical assistance for the dual purpose of enhancing the commitment of local governments to policy goals and increasing their capacity to act. The logic behind this is that higher-level governments know that action must be taken to meet specific policy goals, but local governments are best situated for determining how to achieve those goals. Local governments are told to think seriously about the problems and their solutions following prescribed planning processes, but the specific actions are left to the local governments to determine.

Federal and state policies governing hazards are evolving in the direction of more intergovernmental cooperation. The primary federal example of this approach is the Coastal Zone Management Act of 1972, although it has been amended over the years to add greater prescription about state actions, including attention to hazards (Godschalk, 1992). Also, after two decades of experience with the limited regulatory mandate of the National Flood Insurance Program, the pendulum is swinging the other way as federal monitoring of local floodplain management is reduced in favor of local self-evaluation, as states assume stronger technical assistance roles, and as incentives through the Community Rating System are offered for enhanced local floodplain management programs. A number of states which have recently enacted or revised comprehensive-planning mandates have sought a more cooperative approach to local land use planning and development management. These states have established collaborative partnerships in which the states mandate local government planning processes with varied degrees of prescription about the form and content of the plans. The state mandates vary in the extent of oversight by state governments and the degree of coercion that is employed to get local governments to comply with planning requirements.

Better examples of this approach can be found in other countries. The cooperative approach to hazards management is a central part of environmental management in New Zealand and hazard-management in New South Wales, Australia. New Zealand has received worldwide attention for its reform of resource and environmental management and its vision of integrated, "effects-based" environmental management. Although less comprehensive, the intergovernmental regime in New South Wales provides a noteworthy example of a shift from a heavy hand to a more flexible approach to help local governments cope with flood hazards.

A study by May and his colleagues (1996) of the experience of local governments with cooperative planning programs provides a number of relevant findings. The authors compared what they labeled as a coercive regime, found more typically in the United States (such as Florida's growth management program), with cooperative regimes such as the programs in New Zealand and New South Wales. Both coercive and cooperative forms of intergovernmental policies present dilemmas. The dilemma for coercive regimes is that in bringing about procedural compliance by lower-level governments and forcing them to pursue state-initiated innovations, they may straightjacket local governments that want to develop their own innovative solutions to environmental problems. Coercive intergovernmental regimes can foster a "cookie-cutter" compliance mentality in meeting state-defined deadlines and objectives that tends to emphasize procedural over substantive compliance. The dilemma for cooperative regimes is that, although they foster local ownership of environmental management programs, they suffer from gaps in compliance because of the reluctance of some local governments to participate or follow the prescribed planning process. Local ownership is important because it provides the constituency base for fashioning acceptable land use and development rules, thereby enhancing prospects for implementation and effective environmental management. But gaps in compliance are troublesome, both in themselves and because they are difficult to close.

This research, along with findings summarized earlier from analysis of planning mandates in the United States, provides promising directions for developing more effective federal or state frameworks for governing land use and development in hazardous areas. As noted by Deyle (1995), successful collaboration depends on a confluence of favorable environmental conditions, political will and leadership, an adequate infusion of resources, an effective convener, and sufficient time for a collaborative relationship to mature through experience, mutual learning, and the development of trust between the collaborators. Perhaps the greatest challenge in natural hazards management will be forging a common view of the appropriate policy goals between local governments and higher levels of government.

CONCLUSIONS

The variety of federal and state policies and programs attempting to influence local decisions about land use and development in hazard-prone areas has resulted in an ad hoc patchwork of land use governance. Federal efforts are limited in focus, uneven across different natural hazards, and in some cases undermine the objectives of limiting exposure of life, property, and environmental resources to damage from development of hazardous areas. States employ a greater diversity of management strategies, including greater use of direct regulation of private sector land use as well as an array of incentives, investment policies, and planning and regulatory mandates. The state and federal patchwork of programs is characterized by gaps in coverage and compliance and by missed opportunities for using land use solutions to mitigate hazards.

Another consequence of the land use planning patchwork is the fostering of uneasy intergovernmental relationships. Simply put, local governments do not like the way in which the federal and state governments have attempted to encourage or force their attention to problems posed by natural hazards (or to other environmental issues). Many local officials perceive federal and state environmental mandates as overly prescriptive and coercive. This is a recurring theme of major studies undertaken in the past decade by the U.S. Advisory Commission on Intergovernmental Relations (1993) and the General Accounting Office (U.S. Congress, GAO, 1990a, 1995). Local governments complain about the failure of higher-level governments to fund implementation, about the lack of flexibility in the required actions, and about taking the political blame for infringement on property rights.

The clear implications of the uneasy intergovernmental relationships are a divergence in policy goals and a lack of trust among different levels of government, resulting in a reluctance by local officials to fully implement federal or state policies. The local responses to many federal and state programs are often token, half-hearted compliance; delay in hopes that the higher-level programs will be altered; or in some instances outright refusal to participate. For example, when the federal government sought more state involvement in the National Flood Insurance Program, formerly a federal-to-local program, the states varied greatly in their willingness to take part (May and Williams, 1986). Illinois and other states established strong state programs, but most states mounted only a token effort, and some, such as Oregon, initially refused to establish a state program.

Closely related to the goal divergence and mistrust among different levels of government is the confusion created by inconsistencies among different policies. This occurs at two levels. First is the inconsistency among policies promulgated within a given level of government, notably the long-standing conflict between policies that promote development (e.g., flood control projects and flood insurance) and policies that seek to limit exposure to risk (e.g., regulation of wetlands and development on coastal barriers). The second level of inconsistency is that between state and federal provisions; which provisions take precedence has been an important issue in court decisions about environmental policies. When conflicts exist between regulatory statutes, the federal statute generally prevails under the interstate commerce clause of the U.S. Constitution. However, where federal programs rely on incentives and inducements, there is the potential for divergent state policies to undermine federal initiatives. For instance, although the federal Coastal Barrier Resources Act constrains use of federal funds for growth-inducing infrastructure on coastal barrier islands, states remain free to cover the shortfall with state funds.

Fragmentation of programs and inconsistencies among layers of government are not unique to natural hazards policy making within the American system. Indeed, these are common features of our federal system of government. Yet, the complexities that are introduced by these features are perhaps as strong in relation to natural hazards as in any area of policy.

PART TWO

The Land Use Planning Alternative

CHAPTER FOUR

Integrating Hazard Mitigation and Local Land Use Planning

DAVID R. GODSCHALK, EDWARD J. KAISER, AND PHILIP R. BERKE

HAZARD MITIGATION AND LAND use planning share a *future orientation*. They are concerned with anticipating tomorrow's needs, rather than responding to yesterday's problems. Both are *proactive* rather than reactive. Both seek to gear immediate actions to longer-term goals and objectives. Together they can be powerful tools for reducing the costs of disasters and increasing the sustainability of communities.

This chapter examines the role of local land use planning in mitigating the threats posed by natural hazards. While we acknowledge the importance of federal and state planning and mitigation policies, we take a "bottom-up" local view of the intergovernmental system for hazard mitigation rather than a "top-down" state and federal view. We explore local land use planning's basic powers and authority; the creative process of integrating land use planning with mitigation; the critical choices to be made among levels of stakeholder participation, plan components, plan types, and mitigation strategies; and principles for crafting high-quality mitigation plans.

Integrating hazard mitigation and local land use planning is not a new idea. Hazard specialists have long argued that it makes sense to prevent development in hazardous areas, to design structures to withstand stresses imposed by hazards, and to build public facilities, such as roads and bridges, to stand up under hazard forces while allowing for evacuation of threatened populations (Baker and McPhee, 1975; Berke and Beatley, 1992b; Burby and French et al., 1985; Godschalk et al., 1989; and White, 1945). However, the general public and locally elected officials tend to minimize the importance of discouraging development in hazardous areas. In fact, localities sometimes adopt public policies that unwittingly encourage such development.

What is new about our approach is its linkage of land use planning and hazard mitigation with each other and with *sustainable communities*—where people and property are kept out of the way of natural hazards, where the inherently mitigating qualities of natural environmental systems are maintained, and where development is designed to be resilient in the face of natural forces (as discussed in more detail in Chapter 8, and in Berke, 1995, and Munasinghe and Clarke, 1995). We believe that the time is ripe for such an approach, given the current awareness of the unsustainability of contemporary land use and urban development practices. In the United States, the staggering costs to taxpayers of recent natural disasters such as Hurricane Andrew ($1.64 billion) and the Northridge earthquake ($3.32 billion) have increased public awareness of the need to act beforehand if urban areas are to protect themselves against future disaster threats. To do this successfully requires an understanding of the way in which the values underlying the concept of sustainability can be integrated into the practice of local land use and hazard mitigation planning.

In the pre-disaster period, sustainability values seek to avoid saddling future generations with sprawling, wasteful land use patterns that not only reduce the social livability and economic viability of communities, but also undermine the ability of the natural environment to absorb hazard forces and expose people to significant hazard risks. In the post-disaster period, sustainability values seek opportunities to relocate land use out of hazard areas and rebuild damaged homes and infrastructure in more resilient ways instead of replicating brittle and unsustainable development practices.

PLANNING FOR SUSTAINABILITY

Mitigation planning combines technical analysis and community participation to enable communities to choose wisely between alternative strategies for managing change, and thereby to achieve long-term community sustainability. To that end, the planning program is intended to produce a plan for avoiding or mitigating harm from natural disasters, and for recovering from disasters. The plan details strategies for implementing policies, regulations, capital improvements, and taxing and pricing mechanisms that recognize the economic and equity aspects of sustainability. The planning process also generates valuable information about long-term threats posed by natural hazards to the safety and viability of human development and environmental resources; and it assists with problem solving before, during, and after disasters.

Planning Applied to Mitigation: In Concept

A mitigation plan is a statement of intent. It states aspirations, principles of action, and often specific courses of action the community intends to follow to achieve those aspirations. It is formulated through a systematic process involving a broad representation of community citizens, stakeholders, and officials, and it commits the community to a course of action designed to accomplish considered goals—to reduce losses to private property or to reduce vulnerability of lifeline facilities. Making a plan serves several purposes, some obvious and some not so obvious.

First, the planning process gives a community the opportunity to consider community issues in a systematic and comprehensive manner. It is a practical way to raise, study, and debate issues, and to articulate community goals. In the initial stages, the planning process requires much input of information, such as data on hazards, their geographic variation, and the degree of threat they pose (see Chapter 5).

Second, the process demonstrates the connection, the rational nexus, between public interests and proposed policies or programs. For example, the plan can document the likelihood of property damage if development is permitted in high-hazard areas. In other words, the planning process provides a rational basis for action that is necessary for both political and legal defensibility.

Third, in the course of generating the information necessary to making decisions, the planning process educates the community, and particularly those with a stake in the outcome of the process, about natural hazards and what is feasible to do about them.

Fourth, after acquiring and assimilating information and debating the issues, the planning process allows participants to reach a consensus on goals and actions, to resolve conflicts over priorities and methods, and to build commitment to elements of the plan.

Fifth, a long-range and comprehensive plan coordinates the multiple issues, goals, and policies of a community. For example, hazard mitigation goals, policies, and programs can be integrated with other community goals, policies, and programs for economic development, environmental quality, community development, housing, and infrastructure programming. This avoids uncoordinated and possibly conflicting policies and actions.

Sixth, the plan documents a community's goals, policies, and programs and communicates them to citizens and interested stakeholders, from developers and property owners to elected officials. For example, a plan can ensure widespread understanding of the way that future development will be directed away from high-hazard areas, such as floodplains. The plan alerts everyone to the community's commitment.

Finally, the plan is a means of implementing policy. It is a reference for elected and appointed officials to use in reaching decisions when considering ordinances, allocating money for capital improvements, or granting permits for new developments. It encourages private developers to follow the adopted hazard mitigation policy in order to expedite their permit applications. It is a guide toward coordinating the community's actions along consistent lines.

The Values Supporting the Concept of Sustainability

Land use planning must deal with three powerful sets of community values associated with land (Kaiser et al., 1995). Social values refer to the weight that people give to land use arrangements as the cultural settings for living their daily lives. Market or economic values express the weight that people give to land as a real estate commodity. Ecological values express the weight that people give to maintaining natural plant and animal communities on the land.

These sets of values also underlie the concept of sustainable communities (Clarke and Munasinghe, 1995). One *social* value that is fundamental to the idea of sustainability is the value it places on participation of at-risk communities in planning to reduce vulnerability. Sometimes those most at risk are the poor, the group most likely to live in hazard areas because they cannot afford better, safer land. To increase awareness of hazards and opportunities for mitigation, planners must communicate and consult with those at risk. This may entail decentralizing some decision making to give neighborhoods more control over the choice of mitigation strategies and to promote responsibility among individuals and organizations.

The basic *economic* value underlying sustainable communities is the desire to preserve and enhance a community's capital assets. These include both natural and man-made capital, which complement each other and increase the community's capability to remain viable in the face of shocks. Cost-effective methods of preventing environmental catastrophes also prevent or reduce losses to productivity that constrain economic growth. Economic resources should support the present population without decreasing the economic opportunities of future generations, that is, they should be characterized by intergenerational equity.

Finally, sustainable communities stress the ecological value of preserving the resilience of biological and physical systems and their ability to adapt to change. This involves limiting environmental degradation from pollution and uncontrolled urban development, as well as maintaining those natural systems, such as wetlands and dunes, that increase a community's ability to absorb external shocks from natural hazards. It involves maintaining biodiversity, which enhances the adaptive capacity of natural systems. The connection between degradation of natural resources and vulnerability to natural catastrophe demands preventive planning.

A Model Linking Land Use Planning, Mitigation, and Sustainability

One way to visualize the linkages between land use planning, mitigation, and sustainability is to think of them as parts of a three-legged stool, as shown in Figure 4-1. The seat of the stool represents mitigation/land use planning. The three legs of the stool are social values, market values, and ecological values. Each of these must be balanced to support long-term sustainability, the ultimate goal of the process. A community mitigation/land use planning effort cannot tilt its goals in only one direction without endangering the balance and viability of the community's survival. The three-legged stool metaphor illustrates the dependence of the planning program upon a clearly stated and balanced set of social, market, and ecological values. (See Kaiser et al., 1995, Chapter 2, for a more detailed discussion of this metaphor.)

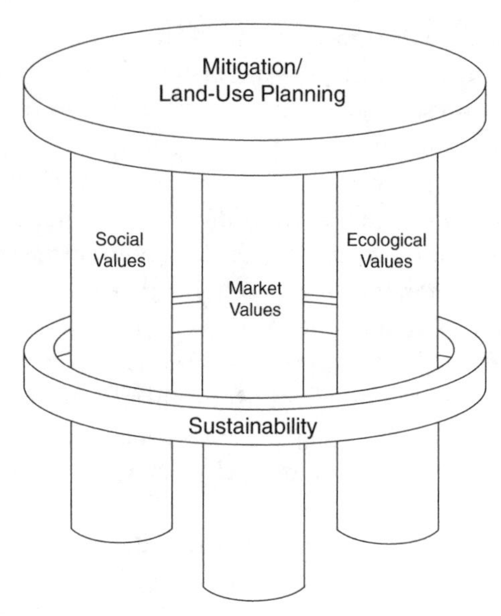

FIGURE 4-1 Community values, mitigation/land use planning, and sustainability. Source: Kaiser et al., 1995.

UNLEASHING THE POWER OF LOCAL LAND USE PLANNING

Local actions can have a powerful impact on losses due to natural disasters; for example, actions taken after the 1993 floods in Missouri reduced the damages from the floods that followed in 1995. Arnold, Missouri, acquired 85 residential structures, 2 commercial structures, and 143 mobile home pads in a buyout program following the 1993 floods (Kaiser and Goebel, 1996). Local officials stated that the impact of the buyout was tremendous. When the fourth largest flood in the city's history struck in 1995, most of the flooded areas had already been purchased, and those residents did not face evacuation and the need for disaster relief.

According to the 1995 FEMA national mitigation strategy, "all mitigation is local." While mitigation against hazards ultimately requires local action, it is important to acknowledge that local action typically occurs in an intergovernmental framework of federal and state policies and programs aimed at empowering and motivating local governments to build mitigation into their plans and actions. In the Arnold, Missouri, example, the funds for the local buyout came from federal disaster assistance and community development programs.

As discussed in Chapter 3, authority, requirements, and incentives for planning and mitigation come down to local governments from the top of the intergovernmental framework. The federal government, under the Stafford Act and other laws and programs, supplies disaster mitigation funding, policy and program guidance, and technical assistance to states and localities. State governments, under their constitutional powers, enact laws transferring authority to local governments for planning and regulating land use, building design, and environmental protection. Local governments make use of these programs, assistance funds, and laws when formulating hazard mitigation plans.

In hazard mitigation, as in other areas, the powers of local governments vary according to their goals and methods:

- *Planning power*—To gain community agreement on a land use plan, local governments can educate, persuade, coordinate, encourage participation and consensus, and offer a vision of the future.
- *Regulatory power*—To direct and manage community development in order to achieve desirable land use patterns and mitigate natural hazards, local governments can use tools of zoning, subdivision regulations, building codes, sanitation codes, design standards, urban growth boundaries, wetland and floodplain regulations, and the like.
- *Spending power*—To control public expenditures to achieve community objectives such as concurrency of infrastructure provision with growth or restricting provision of infrastructure within hazard areas, local governments can use capital improvement programs and budgets.
- *Taxing power*—To support community programs such as infrastructure building and hazard mitigation, local governments can use such tools as special taxing districts and preferential assessment for agriculture and open-space uses.
- *Acquisition power*—To gain public control over lands such as hazard areas, local governments can make use of the right of eminent domain, can purchase development rights, and can accept dedication of conservation easements.

Thus, local governments have considerable power to plan and regulate land use and development. (See Chapter 6 for further discussion of the applications of this power.) But while there are impressive examples of integrating mitigation into local land use practices, overall the level of local response to hazards has been limited. [Berke and Beatley (1992b), Burby and French et al. (1985), and Godschalk et al. (1989) describe local land use practices that have been developed to reduce losses from earthquakes, floods, and hurricanes, respectively.] The lack of effective mitigation programs is not surprising given such obstacles as minimal public interest in natural hazards and the difficulty of operating programs in an intergovernmental setting.

May and Williams (1986) characterize the difficulties of implementing disaster policy under shared governance as a basic dilemma for hazard mitigation. On one hand, federal and state officials have an important stake in promoting hazard mitigation given the escalating disaster reconstruction costs they face (Burby et al., 1991). On the other hand, local governments and property owners in hazard areas are reluctant partners in implementing shared government risk reduction programs, given the lack of a constituency for hazard mitigation.

One way to resolve this policy implementation dilemma is to include mitigation planning in the more commonly used land use planning process, and to design hazard mitigation strategies that also achieve other community goals, such as protection of natural resources and provision of recreation areas and open spaces. The objective of this chapter is to lay out such an integrated approach.

BUILDING HAZARD MITIGATION INTO LAND USE PLANNING

In many communities, hazard mitigation plans are prepared by emergency management staff members and are not tied to comprehensive plans. The products are separate, stand-alone hazard mitigation plans. Later in this chapter, we discuss stand-alone mitigation plans and the choices they involve in more detail. First, however, we describe an integrated approach that fits hazard mitigation into a more comprehensive community land use planning context. When possible, such an integrated approach is preferable to stand-alone mitigation planning.

Context and timing are important factors distinguishing stand-alone mitigation planning from mitigation planning that is integrated with comprehensive planning. Stand-alone mitigation plans may be prepared when communities have not previously adopted comprehensive plans or their comprehensive plans are hopelessly out of date. Often, though not always, stand-alone mitigation plans are prepared in the aftermath of a recent disaster. Thus, they tend to focus on averting a future disaster like the one just experienced by the community. The sense of crisis leads to a call for immediate action. Sometimes called the "window of opportunity," this post-disaster period can break down resistance to mitigation and encourage innovative policy proposals. However, the pitfalls of such planning include the assumption that the next disaster will be a clone of the last disaster ("fighting the last war"), the tendency to make quick plans without a sufficient hazard information database, and the failure to integrate mitigation plans with other related community plans, such as land use and infrastructure. These pitfalls can be overcome by building mitigation planning into continuing community land use planning.

Land use planning is an ongoing process. (For an overview of this process, see Kaiser et al., 1995.) Plans are usually prepared, implemented, and revised on a five- to ten-year cycle. A community planning board made up of citizens works with the professionals on the planning staff to prepare the draft plans, guide them through a public participation process, and craft their final recommendations to the elected officials who make the ultimate decisions about content, and then adopt the plans. The plans produced incorporate both objective technical information and analyses, and more subjective value perspectives and stakeholder preferences. Hazard mitigation may be one goal of a land use plan, along with other community goals such as coordinating future growth with infrastructure capacity or protecting fragile natural resources. (See Federal Emergency Management Agency, 1996, for an overview of the floodplain management planning process.)

The purpose of the planning process is to generate the needed information and specify the goals, policies, and evaluation criteria of a mitigation/land use program that can ensure sustainable communities where people and property are protected from natural disasters. Steps in the planning process include:

1. *Generating planning intelligence*, which defines the problem and serves as the factual basis of the plan. It usually consists of information on local land use patterns and on community vulnerability to hazards by type, location, and intensity (a hazard assessment); and an assessment of the feasibility and effectiveness of possible solutions (a capability analysis). While intelligence collection and interpretation primarily involve technical activities, disseminating information to the community through a public awareness and education program is essential to building the understanding of, and incentive for, a collaborative mitigation/land use implementation strategy.

2. *Setting goals and objectives*, which describe the plan's benchmarks for achievement derived from the citizen participation process (e.g., targets for reduction in vulnerability; increases in residential land needed to accommodate long-range projected population increases while meeting environmental quality standards). In practice, planners draft the initial goals and objectives statements, but these are modified through various public involvement techniques, such as visioning (Nelessen, 1994), goal setting, and consensus building (Godschalk et al., 1994; Innes, 1992).
3. *Adopting policies and programs*, which lay out the actions required to achieve the plan's objectives (e.g., establishment of a land acquisition program for threatened property in the hazard area; creation of a conservation easement program to protect farmland). Planners analyze alternative policies and programs in terms of anticipated effectiveness, efficiency, equity, and feasibility in achieving the selected goals and objectives. Citizens and stakeholders debate the desirability of various policies and programs from the standpoint of their values and engage in conflict resolution processes, such as facilitation and mediation, as necessary to reach agreement on disputed proposals.
4. *Monitoring, evaluation, and revision*, which seek to adapt the plan to community change (e.g., assessing the effectiveness of a program to acquire land in hazard areas against the plan's objectives and other possible mitigation strategies in light of recent disaster experience; measuring the impacts of a cluster development ordinance on stormwater run-off). Planners track the progress of implementation and evaluate its success through ongoing impact analyses. Citizens and stakeholders take part in evaluation procedures, both at the specified five- to ten-year intervals for formal revision of the plan, and when an event such as a disaster or a major change in the community's economic base necessitates a revision. Consensus building comes into play again when necessary to gain agreement on desired modifications to the plan.

Some see community involvement as a separate activity, done at the beginning or end, and segregated from the other technical and decision-making steps in the planning process. We see it as a parallel, integrated, and continuous part of the planning process. Table 4-1 illustrates the parallel technical and community involvement activities in formulating, debating, and approving plan components during an integrated mitigation/land use planning process.

TABLE 4-1 An Integrated Mitigation/Land Use Planning Process

Planning Step	Technical Activities	Community Involvement
Intelligence	Land use projections Hazard assessment Capability analysis	Awareness program Information dissemination
Goals/objectives	Vulnerability reduction Environmental quality Population accommodation	Visioning/goal setting Consensus building
Policies/programs	Hazard avoidance Risk reduction Growth management	Policy debate Conflict resolution Plan adoption
Monitoring/evaluation	Plan/program performance Mitigation effectiveness Impact measurement	Plan evaluation/revision Agreement on modifications

DESIGNING A PLANNING APPROACH: FOUR AREAS OF CHOICE

There is no single model for designing a hazard mitigation planning approach that ties together the necessary technical and community involvement activities. Instead, the planner and the community must choose among alternatives in four areas. They must choose what methods to employ to involve the community in the planning process; which components of a plan to include and emphasize; what type of plan is most appropriate for the specific situation; and perhaps most importantly, what mitigation strategy to employ. Because of the number of possible alternatives, this is a process of thoughtfully selecting and adapting the alternatives that best serve community needs rather than mechanically following a recipe in a planning cookbook. The combination of choices from among alternatives within each of these four areas comprises the mitigation planning approach. This is a custom-designed planning approach in that choices are mixed and matched, and adapted. The variations in approaches are rich and still evolving in practice, as they should be.

In this section we describe alternatives available for each of the four dimensions—stakeholder participation process, plan components, plan type, and mitigation strategy—and illustrate them from contemporary mitigation planning practice and writing.

Choice One: The Approach to Stakeholder Participation

Enlisting community help and support in formulating a hazard mitigation plan and seeing mitigation measures adopted is the essential first step in planning. Local elected officials are unlikely to vote for mitigation measures that are either highly controversial or about which no one seems to care. Thus, to build public involvement it is first necessary to *promote awareness* by disseminating the findings of a local hazards analysis to the community at large, as well as to specific groups and individuals affected by hazards and mitigation proposals. The awareness efforts can include media campaigns, public school information kits, homeowner and builder/developer seminars, community organization speaker series, and similar public information approaches aimed at informing and motivating the community to address natural hazards.

Once public awareness of natural hazards has grown, a more collaborative involvement of the community is needed to *set mitigation goals and objectives*. This process combines technical planning activities such as preparing alternative goal statements, public participation activities such as taking part in community visioning or goal setting, and political activities such as building consensus among stakeholders and elected officials over adopted statements of goals and objectives. Choices regarding goal and objective setting include both *who* to involve and *how* to involve them. Participants can be viewed in terms of their influence and their interest in mitigation. Figure 4-2 illustrates this concept, with decision makers and direct stakeholders at the center, surrounded by layers of community organization leaders, watchdog group leaders, and finally the general public. The most intense involvement takes place closer to the center.

A number of participation models are available (see "Designing a Mutually Acceptable Process" in Godschalk et al., 1994). The most common one consists of a central planning committee with several task groups or subcommittees, organized by geographic hazard area or by mitigation strategy, such as buyout, for example. Another possibility is a large open community mitigation planning conference followed by smaller task groups that report back to the conference. A third possibility is a planning committee or board, either already existing or set up specifically for mitigation, which invites public participation and comment through public hearings and forums. If mitigation is highly controversial, mitigation planning may be accomplished through a dispute reso-

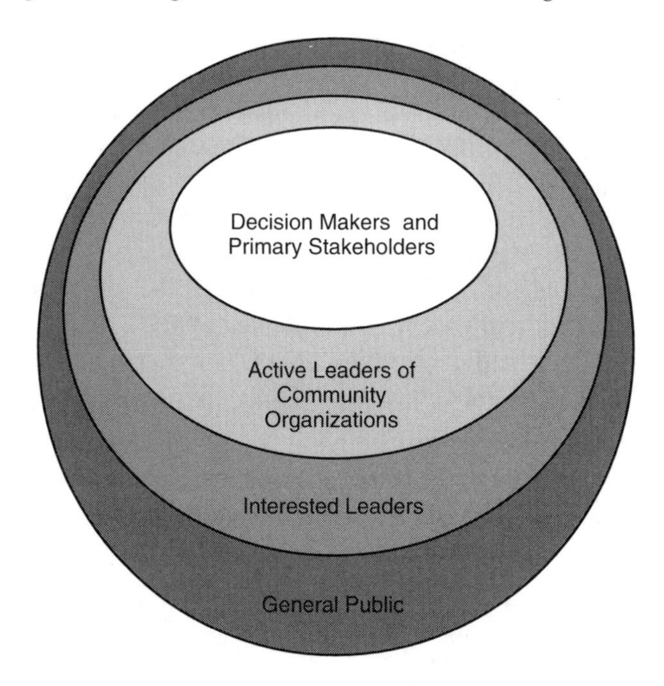

FIGURE 4-2 Levels of participation in mitigation planning. Adapted from Godschalk et al., 1994, Figure 9.

lution process, perhaps involving a facilitator or mediator. The exact form of participation is probably less important than ensuring that all affected stakeholder groups have an opportunity for genuine involvement in the process and are satisfied with its fairness.

Once there is agreement on goals and objectives, the community can *select its implementation policies and programs*. Typically, this is a continuation of the previous process, using the same stakeholders and participation techniques. However, it also is likely to be more contentious, as individual stakeholders debate the merits of policies that have a direct effect on their values. And it may involve more direct attention from elected officials who will be called upon to act on the final implementation agenda by adopting new regulations or approving new public expenditures. Thus, this stage may well call for negotiation among the stakeholders, with the use of dispute resolution techniques, such as facilitation and mediation by neutral third parties, to enhance agreements on adopted plans. This step concludes with the official adoption of the final plan.

Over time, it is necessary to *monitor and evaluate the adopted plans* to determine their effectiveness in meeting community goals for mitigation. During this step, which may take place on a regular schedule (such as every five years in the case of a land use plan) or on an as-needed schedule (such as following a major disaster), the adopted plan is scrutinized and revised. Public participation here can involve reconvening the groups that took part in preparing the plan to review staff analyses, evaluation reports, and potential revisions. Again, a consensus-building effort may be required, particularly if the adopted plan contains clear weaknesses and the evaluation calls for more rigorous mitigation actions that more directly affect important stakeholders.

Choice Two: The Emphasis Within Each Plan Component

As discussed earlier in this chapter, we suggested that the more or less essential parts of a plan are an *intelligence* component, a *goals* component, and an *action or recommendations* component. An *evaluation* component is also desirable.

The Intelligence Component

The intelligence component defines the problem and provides a justification for the policies and courses of action recommended later in the plan. It contains a description of existing and emerging hazardous conditions (a hazard assessment) and factors that bear on feasibility and effectiveness of solutions (a capability analysis).

The *hazard assessment* can be approached in several ways, which are listed below in order of increasing depth and sophistication of analysis. (These approaches are discussed in Chapter 5 in more detail.) The further along in the following order of analytic sophistication in the hazards analysis, the better the factual underpinnings for the plan.

1. Hazard identification locates hazardous areas, often estimates the probability of hazardous events of various magnitudes, and sometimes assesses the separate characteristics of the hazards (e.g., for hurricanes: wind, high water, and wave action).
2. Vulnerability assessment estimates the number of people exposed to hazards (including special populations such as the elderly, hospitalized, disabled, and concentrated populations such as children

in schools), the property exposed, and the critical facilities exposed (such as medical care facilities, bridges, sewage treatment and water pumping and treatment plants, power plants, and police and fire stations).

3. Risk assessment estimates the probable degree of injury and property damage in a given area over a specific time interval.

Sometimes hazard assessments at any of the three levels include analysis of evacuation and shelter demand and capacity for hazards of various magnitudes. Assessments also sometimes include analysis of danger from hazardous facilities (e.g., nuclear plant, chemical plant), and other sources of hazardous materials in the wake of natural disasters of various magnitudes. A good hazard analysis should include assessment of likely future conditions (e.g., projected urban growth) as well as existing situations and past events.

Beyond the *hazard assessment*, the intelligence component should also include a *capability analysis*, describing and assessing the effectiveness of the present hazard management policies and programs. It should also describe other policies and practices (e.g., water and sewer extension policies) that inadvertently are increasing vulnerability to natural hazards. In addition, a capability analysis should assess opportunities and obstacles that are shaping the community's capacities to address hazard mitigation, including statutory authority, rules, and assistance provided by state government.

The Goals and Objectives Component

This component constitutes a statement of community values as a basis for the policies and actions to be recommended in the plan. Possible choices among goals include the following, gleaned from a review of numerous state and local mitigation plans:

- protect the safety of the population
- reduce private property loss
- reduce damage to public property
- reduce government liability
- reduce vulnerability of lifeline facilities (such as hospitals, bridges, power plants)
- minimize fiscal impacts of disasters
- minimize disruption of the economy and social networks
- distribute hazard management costs equitably

- reduce impacts of natural hazards on environmental quality (e.g., water quality, natural areas)
- achieve cost effectiveness in mitigation strategies and actions selected
- achieve a sustainable economy, natural environment, built environment, and social community.

Another way to think about goals is to classify them by their source. Some goals reflect the particular community's concerns and aspirations, which should be elicited in a stakeholder participation process. Some goals, however, may be mandated by state and federal policy. Other goals derive from concepts of the public interest such as efficiency, effectiveness, equity, quality of life, environmental quality, protection of constitutional rights, freedom of choice, and feasibility of proposed actions; the planner has the responsibility of introducing such goals for consideration. Yet other goals are in the form of needs that are revealed in the intelligence component of the plan. For example, there may be a goal to provide for a forecasted demand for evacuation or shelter.

The Action Recommendations Component

The third and most essential component of a plan consists of the policies and/or program of actions to which policymakers are committing themselves. Stronger plans include more specific recommendations for action to achieve the desired goals and objectives. Implementation may include how the proposed policies and actions will be initiated, administered and enforced, and modified; who is responsible for the implementation actions; how the resources required for implementation will be provided; and the timetable for the implementation steps. This component of the plan sometimes includes systematic assessments of recommended policies and actions, including their estimated costs and benefits, and their advantages and disadvantages. The content for this component of the plan is discussed within sections on plan types and mitigation strategies below and in Chapter 6.

The Monitoring, Evaluation, and Revision Component

Ideally, a plan should propose how its hazard assessments and its recommended policies and actions will be monitored and evaluated, and how the plan and mitigation strategy, or its implementation, will be adjusted accordingly over time. This section of the plan should include provision for monitoring the hazards, monitoring implementation progress and problems (including changes in obstacles and opportunities), and assessing successes and failures of policies and programs (i.e., effectiveness). The assessment should include updating baseline data (changing conditions) as well as collecting new data measuring achievement of goals and objectives (i.e., measuring progress). Provision for systematic and timely monitoring, evaluation, and updating of the plan and mitigation policies and programs is sometimes incorporated into the recommendations section rather than addressed in a separate section of the plan. Wherever it is placed, monitoring, evaluation, and revision must not be overlooked if the plan is to remain a meaningful guide to action, because hazards, vulnerability, and risk are dynamic and change over time.

Choice Three: The Type of Plan

While plans vary in their styles, levels of specificity, substantive emphases, and formats, it is possible to classify plans into several types. Not all types of plans are equally suitable for hazard mitigation in all circumstances; the planner and community have options in selecting an approach to fit the preferences of elected officials and other stakeholders, the opportunities presented by state policies and related local capabilities, and the community's other planning efforts. Nor are the types mutually exclusive; the planner may combine features of several types into a hybrid suitable for the specific situation.

The first fundamental choice is whether the mitigation plan will be a separate, stand-alone plan focusing on hazards or will be a part of a more comprehensive community plan. We recommend incorporating mitigation into a comprehensive plan in most situations because that plan normally already has standing in the community as a policy guide, and the comprehensive plan encourages integration of mitigation goals and programs with other ongoing community goals and programs. The danger in making hazard mitigation just one element in a comprehensive plan is that mitigation can become lost in the press of community issues. It might be more effective to write a stand-alone mitigation plan in some situations—for example, when a community has no comprehensive plan or only a weak or out-of-date comprehensive plan, or when the hazard mitigation issue is high on the community agenda, as after a hazardous event, and there is a special opportunity to forge a commitment to a

TABLE 4-2 Choices Between Types of Plans

First-Level Choices	Comprehensive Plan	*versus*	Separate, Stand-Alone Mitigation Plan
	⇓		⇓
Second-Level Choices	Land Classification		General Policy vs. Specific Actions
Land Use Design			Specific Location vs. Community-Wide
	Verbal Policy Land Use Management		Special Hazard vs. All Hazards

mitigation strategy. It is usually possible to integrate a stand-alone plan into a comprehensive plan at a later date.

If integration into a comprehensive plan is chosen, then a second-level choice is whether the plan will be structured as a land classification plan, a future land use design, a verbal policy plan, a land use management plan, or a hybrid of these types. If a stand-alone plan is chosen, then second-level choices are whether to offer general policy guidelines or specify a program of more explicit actions, or do both; whether to focus on explicitly defined hazard-prone areas (like floodplains) or take a more community-wide approach; and whether to deal with a single hazard or to take an all-hazards approach. Table 4-2 summarizes these dimensions of plan type, which are discussed in the following sections.

Making Hazard Mitigation Part of a Comprehensive Land Use Plan

A community's comprehensive land use plan is a broad-based strategy for managing urban change. It addresses the physical development and redevelopment of the community in the intermediate to long-range future. Depending on how comprehensive its approach is, the plan may include sections on public capital improvements, transportation, environmental quality, housing and community development, and historic preservation, in addition to land use and development. In fact, in some states such as California and Florida, and in coastal regions of some other states like North Carolina, city and county comprehensive plans are required by state law to include a section on natural hazards. (For discussion of requirements for a hazards element within a comprehensive plan, see DeGrove, 1992; California Governor's Office of Planning and Research, 1990; North Carolina, 1996.)

The comprehensive land use plan tends to take a policy approach in contrast to a programmatic approach. That is, it offers general guidelines or policies rather than specifying a course of action to be taken. Comprehensive land use plans take one or a combination of several forms. A *land classification plan* consists of a map of various growth policy districts, each district having its own set of development policies. One or more "critical area" districts are often included, including districts highly vulnerable to hazards. The 1974 North Carolina Coastal Area Management Act requires this type of plan and requires that a hazards element be included (N.C. General Statutes, sec. 113A-100 through 113A-134.3). A second and more traditional type of comprehensive plan is a *future land use design*. This design often includes areas designated for nature conservation and open-space uses, which are often purposely located in areas hazardous to urban development but appropriate to

Lifeline failure—the collapse of a 50-foot section of the upper deck of the San Francisco–Oakland Bay Bridge, the only direct vehicular link between San Francisco and the East Bay. Its closure for repairs forced the cross-Bay commuters to take circuitous routes far to the north or south, or to rely on the BART rapid transit system, which was not damaged (NOAA).

natural processes and recreation. Both the land classification plan and the land use design formats emphasize the mapping of appropriate land uses and policy districts. The *verbal policy plan*, a third type of comprehensive plan, eschews mapped policy or end-state visions and focuses on verbal policy statements. Policy regarding mitigation of natural hazards may be one area that is addressed by a verbal policy plan. A strategic plan is often a verbal policy plan, and most land classification plans and land use design plans include a section or sections of verbal policy.

A fourth type of plan is the *development management plan*, a programmatic plan (unlike the other three types), which lays out a specific set of regulations, public investments and acquisitions, infrastructure extension and service area policies, and preferential taxation and pricing policies.

In actuality many local plans are hybrids, combining policies, land use maps, and management programs within a single document.

Comprehensive plans are particularly appropriate for several hazard mitigation purposes:

- Identifying "critical hazard areas," as well as (1) identifying compatible land use activities for those areas; (2) identifying appropriate development and construction standards and practices for future development in such areas; and (3) identifying policies for retrofitting existing development in such areas.
- Identifying areas less vulnerable to hazards, where development and redevelopment will be encouraged and supported, for example, urban growth boundaries (UGBs) and urban service districts that avoid hazardous areas.
- Suggesting lifeline strategies—avoiding hazardous areas in siting community facilities and water and sewer and transportation infrastructure; and imposing hazard-resistant development standards for lifeline facilities.

Traditionally, comprehensive plans tend to be policy plans—that is, they focus on principles of action to guide whole classes of future decisions by public and private decision makers. They are meant to maintain long-term consistency in the stream of incremental community decisions toward goals. However, in recent years comprehensive plans include more programmatic elements; they take the "land use management" approach and recommend specific actions.

Comprehensive plans can incorporate mitigation planning in two ways. One way is to integrate hazard mitigation, response, and recovery pervasively throughout all of the components of the plan, including the assessment of existing and emerging conditions, the framing of goals and general policies, the formulation of community strategies, and the evaluation of plans and implementation, as discussed above. Another approach is to include a separate chapter on hazard mitigation.

California, for example, since 1972 has required city and county plans to include a separate safety element stressing natural hazards (California Governor's Office of Planning and Research, 1990). A safety element is one of seven elements that must be included in the general plan (defined as a comprehensive, long-term plan for the physical development of the community) that every city and county must draw up. A condensed version of the guidelines for this safety element is shown in Sidebar 4-1.

Adding a separate chapter on hazard mitigation to a comprehensive plan might be regarded as something between fully integrating hazard mitigation throughout the comprehensive plan on the one hand and creating a separate hazards plan on the other hand. We hasten to point out, however, that including a separate component on hazard mitigation does not preclude also integrating mitigation policies into other sections of the plan.

The March 1996 draft of San Francisco's community safety element for its master plan provides an example of a local plan under the California approach (San Francisco Planning Department, 1996a). This plan element includes links to other elements of the master plan as well as other local plans and programs, such as the emergency operations plan and post-disaster hazard mitigation plan required by the state. The community safety element of San Francisco's general plan includes a hazards analysis section with maps that delineate hazard areas; an overall hazard mitigation goal; and six categories of objectives-policies-implementation action clusters.

The draft 1996 Comprehensive Development Master Plan for Dade County, Florida, offers a second example of a local plan that incorporates hazard mitigation under state guidelines. One purpose of the plan's coastal management element is to protect human lives and property from natural disasters, and its goals and policies build on assessments based on Hurricane Andrew's impacts, as well as recommendations from the Governor's Commission on a Sustainable South Florida. The objectives of the plan include maintaining or lowering evacuation times and increasing shelter capacity by 25 percent by the year 2000, as well as orienting county planning, regulatory, and service programs to direct future population concentrations away from coastal high-hazard areas through both pre-disaster mitigation and post-disaster redevelopment.

Nags Head, North Carolina, provides another example of a hazards plan as an element or chapter in a comprehensive plan, in this case the city's land use plan (see Sidebar 4-2).

The Option of a Stand-Alone Plan

Another local mitigation planning option is to create a plan that is separate from the community's comprehensive land use plan. An emergency response management plan is a common example of a stand-alone hazards plan, although it rarely includes attention to mitigation. A recovery and reconstruction plan is more likely to incorporate mitigation strategies. Los Angeles, for example, updated its recovery and reconstruction plan in response to lessons learned from the January 1994

SIDEBAR 4-1

California's Requirements for a Safety Element in General Comprehensive Plans

California requires that the general plan of each city and county include a safety element for the protection of the community from any unreasonable risks associated with: seismically induced surface rupture, ground shaking, ground failure, tsunami, seiche, and dam failure; slope instability leading to mud slides and landslides; subsidence and other geologic hazards; flooding; and wildland and urban fires.

The safety element is regarded as the primary vehicle for relating local planning for safety to city and county land use decisions. A city or county must also establish land use planning policies, standards, and designations based on the criteria set forth in the safety element. Likewise, local decisions related to zoning, subdivisions, entitlement permits, and the like should be tied to the safety element's identification of hazards and hazard abatement provisions. The safety element must also specify hazard reduction design criteria and mitigation measures such as land use regulations and conditions of project approval.

The state's guidelines suggest ideas for data and analyses, including maps showing known seismic and geologic hazards, location of special studies zones, and

Northridge earthquake (Los Angeles, City of 1994). The plan contains policies in nine functional areas and describes action implementation programs associated with each policy. Actions are grouped into three groups: pre-event, short-term post-event, and long-term post-event actions. One of the nine functional areas is "Land use/Re-Use," which is where much, though not all, of the mitigation strategy is spelled out, most of it pre-event action. The plan is comprehensive in the sense that every city agency, office, or department has a defined role in implementing the plan, even the land use policies and actions.

A clearer example of stand-alone hazard mitigation planning is provided by the state of Tennessee. Tennessee's state hazard mitigation plan requires local governments to produce simple local hazard mitigation plans in order to receive post-disaster recovery grants. The State Hazard Mitigation Officer and the state's Local Planning Assistance Office and Tennessee Economic and Community Development program have pub-

---

faults and historic data on seismic activity. They suggest geotechnical evaluation of the potential for displacement, ground shaking, mud slides, landslides, liquefaction, and soil compaction. The guidelines also suggest attention to vulnerability, including identification of hazardous or substandard structures that may be subject to collapse in the event of an earthquake; potential for seismically induced dam failure; potential for flooding (including historical data on frequency and intensity); risk of wildland fires and urban fires; emergency evacuation routes as they relate to known hazards; peak load water supply requirements for fire fighting; and other suggested data gathering and analyses.

The state guidelines suggest goals, policies, principles, and standards, including among other things, development standards on setbacks, density, allowable land uses, and site design; requirements for geotechnical evaluations of development sites; establishment of geologic hazard abatement districts; road widths for evacuation routes; and contingency plans for post-disaster response and reconstruction.

The guidelines also list state agencies that are able to provide information or assistance in preparing the safety element of the community's general plan.

SOURCE: California Governor's Office of Planning and Research, 1990.

lished planning guidelines for local governments (see Tennessee Emergency Management Agency, no date, and Tennessee Local Planning Assistance Office, no date). Sidebar 4-3 describes the Tennessee approach. Another form of stand-alone plan focuses on a particular hazard area, sometimes a small part within the jurisdiction. Thus it is less comprehensive than any of the formats discussed above and focuses on a particular community problem. An example is a floodplain management plan, which might delineate the boundaries of a floodplain regulation overlay district, indicate specific properties to be acquired, propose a levee or replacement bridge, and perhaps indicate the location of a wetland flood storage area.

It is important to point out that even the stand-alone hazard mitigation plan can include land use strategies, as evidenced by the Los Angeles and Tennessee examples, and does not imply an avoidance of land use and development strategies for hazard mitigation. Stand-alone mitigation plans also can (and should) be coordinated with comprehensive land use plans.

SIDEBAR 4-2

Incorporating a Hazards Plan as a Chapter of a Comprehensive Land Use Plan (Nags Head, N.C.)

In the *1990 Land Use Plan for Nags Head, North Carolina*, the chapter on hurricane and coastal storm hazard mitigation is one of 20 chapters dealing with topics such as managing growth, traffic and transportation, water and sewer service, economic development, housing, and the like. However, hazard mitigation also is considered in other chapters, such as traffic and transportation, where hurricane evacuation is addressed as a critical issue. The mitigation chapter refers the reader to other planning documents, such as the 1988 *Comprehensive Hurricane and Storm Mitigation and Reconstruction Plan*, which provides for a reconstruction task force, a post-storm development moratorium, and other ordinances, and the Flood Insurance Rate Maps (FIRMs), which identify the various flood hazard areas.

The mitigation chapter of the Nags Head plan first describes the threats to life and property posed by hurricanes and severe coastal storms, maps the location of storm hazard areas (including flood zones and areas where new inlets might be formed), and estimates the dollar amount of property at risk in each hazard zone. It then states its goal: "to reduce, to the extent possible, future damage from

Choice Four: The Mitigation Strategy

A fourth major area of choice in developing the plan is the choice of the mitigation strategy to be employed by the community. In contrast to choices about planning process, planning components, or type of plan, this is a choice about the substance of the plan's recommendations. A particular mitigation strategy involves choices regarding six variables:

1. taking a coercive approach versus a cooperative approach to influence private sector behavior
2. employing one local governmental power versus another
3. shaping future development versus addressing existing development at risk
4. controlling the hazard versus controlling human behavior
5. taking action before a disaster occurs versus taking action afterward, during recovery
6. going it alone versus taking an intergovernmental and regional approach.

---

hurricanes and severe coastal storms . . . both in advance of such events and . . . during reconstruction" (p. 31).

To achieve this goal, the plan spells out a number of policies (pp. 31-32):

"Pre-Storm Mitigation.

1. The adopted capital improvements program will encourage growth away from the highest storm hazard areas and minimize the extent of public investment at risk.
2. Natural mitigation features of the barrier island, such as dunes and wetlands, will be protected and enhanced.
3. Permanent open space for recreation and other public purposes will be increased, wherever possible, by purchasing land or interests in land in high-hazard areas.
4. Federal, state, and regional policies and programs affecting local mitigation will be identified and influenced to reduce local hazards.
5. The National Flood Insurance Program (NFIP) and FEMA's Community Rating System (CRS) will be actively supported and used to prevent storm and flooding damage.
6. Finger canal construction is opposed and will be prohibited."

continued

Coercive versus cooperative approach

The coercive approach prescribes a required behavior, with sanctions for failure to comply. It gives the target population of developers, landowners, residents, or businesses little choice about whether and how to meet requirements. The coercive approach emphasizes regulations, and a more formal and legalistic implementation style in enforcement. The cooperative approach, sometimes called the collaborative or co-production approach, prescribes goals to be achieved but not the precise means of attaining them. To enhance adherence to policy objectives, the more coercive approaches employ more and heavier sanctions and disincentives; cooperative approaches seek to induce change by offering incentives, removing obstacles, or educating stakeholders, and taking a more flexible approach to regulations and their enforcement. [May et al. (1996) and May and Burby (1996) compare the performance of coercive and cooperative approaches to hazard mitigation.] A hazard plan might

SIDEBAR 4-2 *Continued*

"Post-Storm Reconstruction.

1. All reconstruction will be required to conform to the North Carolina State Building Code.
2. Damaged public facilities, including water lines, will be rebuilt consistent with town practices, policies, regulations, and objectives.
3. Opportunities to purchase land in hazard areas following a storm will be taken.
4. Redevelopment patterns which recognize and utilize natural mitigative features of the coastal environment will be encouraged.
5. Recovery and reconstruction planning will be integrated with broader community goals and objectives, including the opportunity to modify existing development patterns following a hurricane.
6. Redevelopment shall occur only at the intensity permitted in the zoning district at the time."

SOURCE: Nags Head, N.C., City of, 1990.

incorporate both cooperative and coercive strategies in order to address different types of problems or different target groups.

Choosing between governmental powers

As discussed earlier in this chapter, local governments have several types of powers, including the power to plan, the police power, the power to acquire land and property, the power to invest in capital improvements, and the power to tax. Most mitigation strategies will entail the use of more than one power, but will vary according to which power or powers they emphasize, the coordination among the actual devices utilizing these powers, and even the combination of several powers within a device (e.g., incorporating eminent domain with acquisition, including an impact fee or tax as part of a regulation).

Shaping future development versus addressing existing population and property at risk

The plan may choose to focus on future development or on existing development or both, depending on the relationships between hazards and urban land use patterns. The first option implies limiting new private and public development in hazard-prone areas, requiring hazard-proofing of new development, or requiring insurance, for example. Specifically, a plan concerned with future development could: (a) change the types and densities of uses allowed in the hazard area; (b) prescribe the engineering and site-design standards required of new development in the hazard area; (c) specify building design standards to strengthen structures; (d) improve knowledge about the hazard's risk and spatial variation in risk; (e) structurally control the hazard; and (f) protect the community lifeline facilities. The second option, focusing on existing development, would imply requirements or incentives for retrofitting, or acquisition of property at risk and relocation of residents and businesses, for example.

Controlling the environmental hazard versus controlling human behavior

The plan might propose control of the hazard itself in order to reduce vulnerability of populations and property, for example, by constructing or improving levees, stabilizing slopes, building flood control dams, or installing stormwater retention and drainage infrastructure. In contrast, the plan could direct more effort to controlling human behavior by discouraging incompatible use of hazardous areas or by allowing their use but making development less vulnerable to the hazard (e.g., by imposing elevation standards for buildings in floodplains). In some cases, both types of control may be needed.

Taking action before an event versus taking action as part of a recovery program

Pre-event mitigation is, by definition, more anticipatory and preventive, and addresses future development. Post-event mitigation, however, makes use of critical windows of opportunity to rebuild without remaking past mistakes. A good plan addresses both pre- and post-disaster mitigation.

SIDEBAR 4-3

Tennessee's Guidelines for a Local Hazard Mitigation Plan

Tennessee's State Hazard Mitigation Officer and Local Planning Assistance Office have collaborated to provide local governments with guidelines for creating a local hazard mitigation plan. Consistent with FEMA's guidelines for state hazard mitigation plans, the recommended parts of the local plan include the following:

- *A hazard evaluation element to determine the disaster potential.* This component has two parts: (1) an identification of natural hazards for the local jurisdiction including a description of their frequency, probability, magnitude, and distribution (with a historical summary and maps of areas affected); and (2) a description of vulnerability to the identified hazards, to establish who and what is at risk and to what extent (including the populations and value of property at risk), impacts on tax base and public infrastructure, and assessments based on both existing and potential development.
- *A hazard management capability assessment.* This component is intended to develop an understanding of the status of current systems (e.g., laws, standards, programs, and policies) that either reduce or increase vulnerability to hazards, including existing policy that could be effective but for some reason is

Going it alone versus taking an intergovernmental or regional approach

Local governments may choose to coordinate their mitigation strategies with adjacent jurisdictions or a group of jurisdictions within a larger region, or even become part of a statewide strategy. In growth management states, state government may mandate that local plans include mitigation of natural hazards and, further, that such plans exhibit vertical and horizontal consistency. Vertical consistency is a requirement that local plans be consistent with state hazard mitigation goals and policy. Horizontal consistency, sometimes called regional coordination, is a requirement that local mitigation plans be coordinated with those of neighboring governments. In several states, regional planning agencies help local governments attain horizontal consistency by providing regionwide information on hazards and mediating intergovernmental conflicts about how best to address hazards.

---

not. The assessment should classify activities that could be done easily, those that require a change in regulations, and those that require new authority.

- *A statement of goals, objectives, and strategies.* This component should include far-reaching goals, a series of shorter-term objectives that will be used to achieve the goals, and very specific mitigation strategies in the form of policies, standards, programs, and projects that will implement the objectives. Each mitigation strategy includes a statement of the problem, alternative solutions and the preferred action, identification of a lead agency to implement the action, an implementation schedule, the source of funding, and the estimated cost.
- *An implementation, evaluation, and updating procedure.* Implementation includes adoption by appropriate bodies. Evaluation is on a semiannual basis to assess implementation, monitor progress, and observe changes in context, community values, and hazard conditions. Local plans are updated on an annual basis to incorporate new issues that arise during evaluation and to report progress to the State Hazard Mitigation Officer.

Easy-to-use forms and an illustration of a very simple hypothetical hazard mitigation plan are provided in a separate publication.

SOURCE: Tennessee Emergency Management Agency, and Tennessee Local Planning Assistance Office, no dates.

ASSESSING PLAN QUALITY

Every plan brings together a series of choices designed to fit the unique needs of a particular community. Yet there are accepted principles for determining what makes a good plan. It is possible to evaluate plans of whatever type, emphasis, and strategy according to established practice criteria.

This section lists criteria to be considered in producing and evaluating hazard mitigation plans. The criteria are grouped under 12 principles, which state key concerns to be considered in preparing a plan, including clarity of purpose, citizen participation, identification of issues, specification of policies, quality of fact base, integration with other plans and policy, coordination with other community development goals, planning for multiple hazards, organization and presentation, internal consistency, monitoring, and implementation.

The principles are derived from several sources. They are taken from research on the influence of state mandates on comprehensive plans, and on the plans' effectiveness in influencing adoption of hazard mitigation measures (Berke and French, 1994; Berke et al., 1996a; Burby and Dalton, 1994; Burby and May et al., 1997; Dalton and Burby, 1994). The principles are also derived from research on how well plans are integrating the concept of sustainable development in New Zealand (Berke et al., 1996) and in 30 communities in the United States (Berke and Manta, 1997). Another source is evaluations of the effects of these principles of plan quality on mitigation measures adopted by the states under the Stafford Disaster Relief Act (Godschalk et al., 1996).

These principles are not conclusive. They should not be viewed as a checklist, with each principle corresponding to a particular compliance criterion required by state planning mandates. Rather, these principles should provide guidance, with user discretion required as to their application in particular local circumstances. They reflect basic planning concepts and are intended as a starting point, to help planners think systematically about what should be included in a good mitigation plan. Given variation in local purposes and circumstances, there may be differences in the applicability of different principles and criteria. Communities should modify the principles to fit their own needs.

Principles and Criteria for Preparing and Evaluating Mitigation Plans

1. *Clarity of purpose*: the plan should clearly identify and explain the desired mitigation outcomes.
   - 1.1 Is there a clear articulation of desired mitigation outcomes?
   - 1.2 Are mitigation outcomes linked to broader environmental, social, and economic outcomes?
   - 1.3 Is there a clear discussion of how the plan can be used to effect mitigation outcomes (that is, of the connection between public interests and actions, and of methods of implementation)?
   - 1.4 Is there a clear explanation of state and federal legislation enabling or requiring the plan?
   - 1.5 Does the plan clearly explain how legislative provisions are to be implemented?
2. *Citizen participation:* the plan should be based on explicit procedures for involving stakeholders.
   - 2.1 What organizations and individuals were involved in the preparation of the mitigation plan (public officials, elected officials, citizens, representatives of stakeholder groups, consultants)?
   - 2.2 Why were they involved (e.g., selected due to expertise or because they represent a particular constituency, or self-selected volunteers)?
   - 2.3 Are the stakeholders who were involved representative of all groups that are affected by polices and implementation actions proposed?
   - 2.4 What participation techniques were used (e.g., citizen participation program, blue ribbon panel, advisory commission, focus groups, charettes/workshops)?
   - 2.5 To what extent do participation techniques elicit passive stakeholder involvement (e.g., by provision of educational materials, or by citizen review of plans at public hearings) versus active involvement (e.g., through neighborhood organization efforts, or through citizen control of resources and direct public and private investment decisions)?
   - 2.6 Is there a historical account of how stakeholder involvement in the plan is related to prior planning activities?
3. *Issue identification*: the plan should discuss and prioritize significant mitigation issues.
   - 3.1 Are issues related to natural hazards clearly identified?

3.2 Is background information on each issue, including causes and effects, clearly explained?
3.3 Are issues prioritized?
3.4 Is the rationale for prioritizing issues clearly explained?

4. *Policy specification*: the plan should provide specific policies to guide decision making and planning.
   4.1 Are mitigation policies specific enough to be tied to definite actions? (For example, a vague policy would be to reduce flood risk; a specific policy would be to reduce development densities in floodplains.)
   4.2 Are mitigation policies merely suggested (with words like consider, should, may), or are they mandatory (with words like shall, will, require, mandate)?
5. *Fact base*: the plan should be built on a solid database foundation.
   5.1 Are explanations of issues, current conditions, trends, and likely future conditions supported by good quality data?
   5.2 Are explanations of mitigation policy proposals congruent with scientific data? With common knowledge?
   5.3 Are methods and models used for deriving facts cited?
   5.4 Are maps that identify hazardous areas and vulnerable structures included? Do the maps display information that is relevant and comprehensible?
   5.5 Are tables that aggregate data about vulnerable structures and populations included? Are the formats of tables relevant and comprehensible?
   5.6 Are sources given for background information and data?
6. *Policy integration*: the plan should coordinate its actions with those of other relevant agencies.
   6.1 Are plans internal and external to the community mentioned in the plan?
   6.2 How clear is the explanation of the relationship of each mentioned policy/policy instrument to the plan under study?
7. *Linkage with community development*: the plan should link mitigation and community development objectives.
   7.1 Are mitigation goals tied to other publicly supported development goals (e.g., enhancing open space in recreation areas, environmental conservation, more efficient use of public infrastructure investments)?
   7.2 Are the policy instruments (e.g., zoning, building codes) used for carrying out mitigation also used for other community development aims?

8. *Multiple hazard scope*: the plan should deal with all hazards affecting the community.
   - 8.1 Does the plan include mitigation goals and policies that are generic and effective for all the various hazards a community might face?
   - 8.2 Are hazard risk assessments conducted for multiple hazards rather than single hazards?
   - 8.3 Are risks to life and property posed by different hazards compared and prioritized?
9. *Organization and presentation*: the plan should be understandable to a wide range of readers.
   - 9.1 Does it include a table of contents (not just list of chapters)?
   - 9.2 Does it include a glossary of terms and definitions?
   - 9.3 Does it include an executive summary?
   - 9.4 Is there cross-referencing of issues, goals, objectives, and policies?
   - 9.5 Are clear illustrations used (e.g., diagrams, pictures)?
   - 9.6 Is spatial information clearly illustrated on maps?
   - 9.7 Are supporting documents included with the plan (video, CD, GIS, Web-Page)?
10. *Internal consistency*: the plan should ensure consistency among its various goals, objectives, and policies.
    - 10.1 Are mitigation goals comprehensive enough to accommodate strategic issues?
    - 10.2 Are mitigation policies clearly linked to certain goals?
11. *Performance monitoring*: the plan should include indicators for assessing goal achievement.
    - 11.1 Are clearly defined outcomes stated in the plan (number of historic buildings strengthened, transportation systems that accommodate evacuation times within an acceptable number of hours)?
    - 11.2 Are clear indicators of each outcome included? Are development and land use standards included that suggest thresholds not to be exceeded (e.g., allowable development densities in hazard zones)?
    - 11.3 Does the plan identify organizations that are responsible for monitoring and/or providing data for indicators?
12. *Implementation*: the plan should commit the community to carry out its proposed actions.
    - 12.1 Are timelines for implementation identified?
    - 12.2 Does the plan clearly identify organizations that are responsible for implementation policies?
    - 12.3 Does the data identify sources of funding for implementation?

CONCLUSIONS

This chapter reviewed the role of land use planning in mitigating the risks posed by natural hazards. We emphasized the importance of local governments in planning mitigation strategies for good land use in hazardous areas. We argued that the crucial role of land use planning in mitigation must be anchored in a community-based planning process. Although federal and state governments play a crucial role in setting ground rules for mitigation and enabling communities to deal with the complex land use problems involved, our contention is that local governments must be the driving force. We explained why they are best suited to seek mitigation solutions given their basic powers and authorities. We also discussed why local governments are best able to make critical choices about stakeholder participation processes, and about how to craft high-quality plans that are tailored to local needs and capabilities.

The goal of sustainable development underpins hazard mitigation and land use planning. Achieving a pattern of human settlement that can withstand natural hazards should be a central objective of any community land use planning program. Land use patterns that fail to acknowledge the location of high-risk areas like floodplains, steep slopes, and seismic faults are not sustainable. Housing and public infrastructure not built to resist the predictable physical forces of hazards are not sustainable. Sustainable communities minimize exposure to hazards and enhance resilience in the face of natural forces.

To be more sustainable, communities must integrate hazard reduction with other social, economic, and environmental goals. Every new development project and public investment should be evaluated with regard to several criteria for sustainability. Are ecological limits recognized in public and private land use decisions in hazardous areas? Do land use decisions for hazard mitigation accommodate prospects for enduring economic development? Do land use decisions seek an equitable distribution of risks of hazards and benefits of mitigation? How can all people participate meaningfully in formulating land use and mitigation decisions that affect them? These criteria can be added to the criteria discussed in this chapter for evaluating plans. They also represent an opportunity to increase public understanding of land use management and hazard mitigation.

Local planning practice is evolving from a focus on post-disaster recovery and redevelopment to a focus on pre-disaster mitigation linked to land use planning. Increasingly, planners are learning how to demonstrate to citizens and decision makers how much sense it makes to deal with hazards before they become disasters.

CHAPTER FIVE

Hazard Assessment: The Factual Basis for Planning and Mitigation

ROBERT E. DEYLE, STEVEN P. FRENCH,
ROBERT B. OLSHANSKY, AND
ROBERT G. PATERSON

TRADE-OFFS BETWEEN RISKS are inherent to the land use management choices made by property owners, planners, and government officials for areas subject to natural hazards. Prospective property owners choose whether to invest in property and how to use the land they purchase. They make choices about design of the site and the structures built there, construction materials, and insurance. Local officials, with input from land use planners, make comparable choices about where to direct land uses and where to provide public infrastructure and facilities to support private land use. In making such choices, they face trade-offs between risks to public safety and public and personal property on the one hand, and economic benefits that may result from developing hazardous areas on the other. Local officials, as well as state and federal officials, also make choices about how to reduce the risks associated with development in hazardous areas. Choices may be the unintended or unconsidered result of official action, rather than the result of deliberately weighing the benefits and risks of alternatives. Indeed, choices involving risk are frequently made in the face of substantial misunderstandings or uncertainty about the severity or likelihood of adverse consequences.

Sustainable land use cannot be achieved for hazardous areas when decision making is not adequately informed about risk. To make informed choices, local officials and their constituents must know how many people are subject to injury, how many structures can be damaged, and how much infrastructure can be lost, as well as the likelihood that such impacts will occur. They also must understand how changes in land use can affect natural hazards themselves. Hazard assessment is intended to provide the factual basis for estimating the likely costs and benefits of alternative land use scenarios and various strategies for reducing risks. Knowledge of the risks posed by extreme natural events and an understanding of how such knowledge can influence human behavior are, therefore, central to assessing the potential of land use planning and management strategies for achieving safer, more sustainable communities.

This chapter examines the choices that face local officials when they use hazard assessment, and summarizes the state of practice and the state of the art in applying hazard assessment to evaluate and implement alternative land use planning and management strategies. We begin by explaining the basic concepts and levels of hazard assessment and discussing how hazard assessment can be and is used in land use planning and management. We then examine the choices that must be made in applying hazard assessment, including choices about the precision and geographic scale of analysis, temporal perspective and boundaries, and the level of hazard assessment to use. We follow our discussion of these choices with a review of the state of knowledge and information used in hazard assessment, and a detailed summary of the state of practice for floods, earthquakes, landslides, hurricanes and coastal erosion, and wildfire.

This discussion leads to several conclusions. First, the form of hazard assessment that is most often used at present is hazard identification, which is the essential foundation for managing land use in hazardous areas. Communities make considerably less use of vulnerability assessments to predict the impacts of natural hazards, and they rarely use formal risk analysis. Second, the advent of geographical information systems and digitized land use data has opened the way for the development and use of probabilistic risk analysis to predict the impacts of natural hazards and assess the costs and benefits of alternative plans and land use management strategies. Third, the principal obstacles to greater use of these capabilities appear to be: (1) limited knowledge of the probabilities, magnitudes, and locations of some types of extreme natural events; (2) lack of parcel-specific data on relevant attributes of land uses such as the type, design, and construction of buildings; (3) lack of empirically validated data on hazard effects (referred to as damage functions) that are accurate at the building or infrastructure-component level for some natural hazards; (4) lack of professional expertise to incorporate sophisticated risk analysis models into land use decision making; and (5) lack of understanding and confidence in these models by appointed and elected officials.

APPLYING HAZARD ASSESSMENT TO LAND USE PLANNING AND MANAGEMENT

The first step in appreciating the potential utility of hazard assessment is to understand how it is conducted and how it has been used and can be applied to land use planning and management. The terminology of hazard assessment is muddied by inconsistent definitions and usage of key terms such as "hazard," "vulnerability," and "risk." In this chapter we use the following definitions: *Hazard* refers to an extreme natural event that poses risks to human settlements. *Vulnerability* is the susceptibility of human settlements to the harmful impacts of natural hazards. Impacts of concern include injuries and deaths to human populations; damage to personal property, housing, public facilities, equipment, and infrastructure; lost jobs, business earnings, and tax revenues, as well as indirect losses caused by interruption of business and production; and the public costs of planning, preparedness, mitigation, response, and recovery. *Risk* is "the possibility of suffering harm from a hazard" (Cohrssen and Covello, 1989, p. 7).

Hazard assessment can be conducted at three levels of sophistication:

1. *Hazard identification,* which defines the magnitudes (intensities) and associated probabilities (likelihoods) of natural hazard that may pose threats to human interests in specific geographic areas.
2. *Vulnerability assessment,* which characterizes the exposed populations and property and the extent of injury and damage that may result from a natural hazard event of a given intensity in a given area.
3. *Risk analysis,* which incorporates estimates of the probability of

various levels of injury and damage to provide a more complete description of the risk from the full range of possible hazard events in the area.

All three levels of hazard assessment are necessary to realize the full potential of the process. However, each level provides information that is useful in its own right.

Many land use planning and development management applications use information generated from hazard identification or vulnerability assessment rather than full-scale risk analysis. All three levels of hazard assessment may be used not only for future land use planning but also for redevelopment planning and regulation of existing land uses. When applied to future land use, these techniques may be used to identify and avoid potential problems associated with developing hazardous areas. This is the practice of identifying "critical areas" where development should be restricted or regulated by strict performance standards, such as floodplains, seismic fault areas, and areas prone to wildfires.

Where land already has been developed, these techniques may be used to justify the imposition of requirements on existing development, to define areas where such controls are necessary, and to assess the benefits of other means of mitigating hazards. For example, building codes designed to reduce the risk of significant structural damage have been applied retroactively to existing buildings subject to earthquake damage in Los Angeles and Palo Alto, California. Within designated hazard zones, structures that are damaged beyond a specified threshold, such as 50 percent of market or replacement value, are often required to be rebuilt to current codes, which include measures to reduce vulnerability to natural hazards.

Degree of Use of the Three Levels of Hazard Assessment

Table 5-1 presents estimates of the extent to which the three levels of hazard assessment are used in the formulation, design, or justification of various land use planning and management tools. The table was constructed using a modified Delphi technique with the authors of this volume, based on their familiarity with the practice of planning for hazardous areas. Table 5-1 indicates that hazard identification is the most widely used form of hazard assessment. Hazard identification may be directly incorporated into a planning or management tool, as in the designation of hazardous area overlay zones or the specification of hazard setbacks in subdivision ordinances. Vulnerability assessments are used

TABLE 5-1 Use of Hazard Assessment in Land Use Planning and Management

Plans and Implementation Tools	Hazard Identification	Vulnerability Assessment	Risk Analysis
Planning			
Comprehensive plan, hazard component of comprehensive plan, recovery/ reconstruction plan	CP	S	R
Local emergency management plans	CP	CP	R
Development regulations			
Zoning ordinance	S	R	R
Subdivision ordinance	S	R	R
Hazard setback ordinance	CP	R	R
Building standards			
Building code	CP	R	R
Special hazard resistance standards	CP	S	R
Retrofit standards for existing buildings	CP	S	R
Property acquisition			
Acquisition of undeveloped lands	CP	R	R
Acquisition of development rights	CP	R	R
Building relocation	CP	CP	R
Acquisition of damaged buildings	CP	CP	R
Critical and public facilities policies			
Capital improvements programs	CP	R	R
Location requirements for critical facilities	CP	S	R
Location of public facilities and infra-structure in less hazardous areas	CP	R	R
Taxation			
Impact taxes	CP	S	R
Reduced or below-market taxation	CP	R	R
Information dissemination			
Public information program	CP	S	R
Hazard disclosure requirements	CP	S	R

NOTES: CP = Common practice; S = Sometimes used; R = Rarely, if ever, used.

primarily to build support for hazard management policies and programs. Their potential for evaluating the relative benefits of alternative policies has not yet been realized fully. Full-scale risk analyses have not been used extensively, possibly because planners and local officials are less familiar with risk analysis concepts and methods, and because of the relative paucity of land use management tools that are based on risk rather than vulnerability.

Hazard Identification

Hazard identification, the essential foundation upon which all hazard assessment is based, is the process of estimating the geographic extent of the hazard, its intensity, and its probability of occurrence. Geographic extent may vary with the magnitude of the event, as with riverine flooding, but this is not always the case. For instance, stronger hurricanes do not necessarily affect larger geographic areas than hurricanes with higher central pressures and lower wind speeds.

Intensity refers to the damage-generating attributes of a hazard. For example, water depth and velocity are commonly used measures of the intensity of a flood. For hurricanes, intensity typically is characterized with the Saffir/Simpson scale, which is based on wind velocity and storm surge depths (see Table 5-2). The absolute size of an earthquake is given by its Richter magnitude (and other similar magnitude scales), but its effects in specific locations are described by the Modified Mercalli Intensity (MMI) Scale (see Table 5-3). Earthquake intensity is also ascertained by physical measures such as peak ground acceleration (expressed as a decimal fraction of the force of gravity, e.g., 0.4 g), peak velocity, or spectral response, which characterizes the frequency of the energy content of the seismic wave. For wildfires, intensity can be expressed as fire line intensity (a measure of the rate at which a fire releases heat, or the unit length of the fire line); the rate of fire spread (feet per second); and flame length. A more qualitative intensity measure is based on the primary fuels being burned. Three classes are defined: (1) ground or subsurface fires, which are the lowest intensity and most controllable forms of combustion; (2) surface fires, which can be of low or high intensity depending on environmental conditions and fuel characteristics; and (3) crown fires, which are largely uncontrollable through traditional suppression methods and which have various intensity levels of their own.

TABLE 5-2 Saffir/Simpson Hurricane Scale

Storm Category	Wind Speed (mph)	Storm Surge (ft)
1	74–95	4–5
2	96–110	6–8
3	111–130	9–12
4	131–155	13–18
5	> 155	> 18

TABLE 5-3 Modified Mercalli Intensity Scale (Earthquakes)

Intensity	Detectability/Level Impact
I	Detected only by sensitive instruments
II	Felt by a few persons at rest, especially on upper floors
III	Felt noticeably indoors, but not always recognized as a quake
IV	Felt indoors by many, outdoors by a few
V	Felt by most people
VI	Felt by all, many frightened and run outdoors, damage small
VII	Everybody runs outdoors, damage to buildings varies
VIII	Panel walls thrown out of frames, fall of walls and chimneys
IX	Buildings shifted off foundations, cracked, thrown out of plumb
X	Most masonry and framed structures destroyed, ground cracked
XI	New structures still standing, bridges destroyed, ground fissures appear
XII	Damage total, waves seen on ground surface

SOURCE: Jaffe et al., 1981, p. 61.

The likelihood or probability of a hazard occurring usually is calculated on an annual basis, for example, a 10 percent chance of a particular area being struck by a Category 1 hurricane (Saffir/Simpson scale) or a level VI intensity earthquake (Modified Mercalli scale) in a given year. For many hazards, likelihood often is expressed as a recurrence interval, such as a 100-year storm (a storm with a 1 percent annual probability) or a 475-year earthquake (a 0.2 percent annual probability).

Hazard identification typically takes the form of hazard maps. These may be prepared for a single scenario for an extreme event of a specified intensity such as a 100-year flood, or for the composite of several levels of hazard, defined by intensity or probability, such as hurricane storm surge maps. Hazard maps also may indicate areas that are subject to a particular natural hazard over a range of intensities and probabilities, such as the Boulder County (Colorado) Wildfire Hazard Information Mitigation System which uses a hazard model based on geographic information system technology to assign hazard ratings, on a scale of 0 to 10, to neighborhoods, structures, and lots (Boulder County Land Use Department, 1994).

The Flood Insurance Rate Maps produced by the National Flood Insurance Program are the most familiar example of hazard identification (see Figure 5-1). They delineate the floodway and the floodplain for the 100- and 500-year floods and provide contours that depict flood elevations within the 100-year floodplain. Hurricane storm surge maps follow a different approach, depicting the areas subject to storm surge

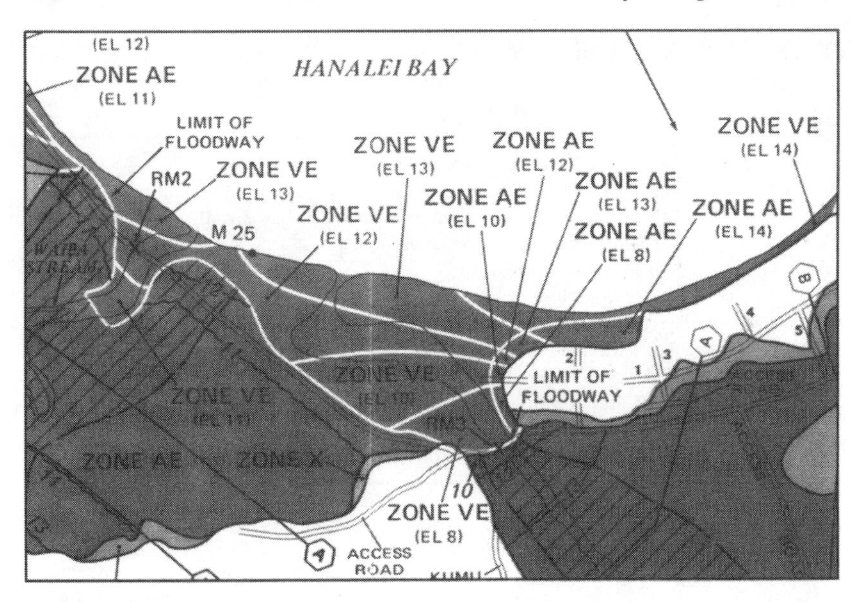

FIGURE 5-1 Reproduction of a portion of a Flood Insurance Rate Map for Kauai County, Hawaii. Key: Zone AE = area of 100-year flood; Zone VE = area of 100-year coastal flood with velocity (wave) action; Zone X = area of undetermined but possible flood; elevations (e.g., EL 13) = Base (100-year) flood elevation in feet [NGVD = National Geodetic Vertical Datum]); A = symbol for cross sections used to compute base flood elevations; M25 = river (coast) mile marker; ▨ = floodway area (the channel of a river or other watercourse plus any adjacent floodplain areas that must be kept free of encroachments so that the 100-year flood discharge can be conveyed without increasing the elevation of the 100-year flood more than a specified amount).

flooding from hurricanes at each of the Saffir/Simpson scale intensities (see Figure 5-2). The probabilities for the individual storm categories vary substantially from one location to another.

Demarcations of earthquake and landslide hazards are not as standardized as flood maps. California's Alquist-Priolo Earthquake Fault Zoning Act (Calif. Public Resources Code, section 2621) defines earthquake fault-zone areas based on proximity to a fault that has been active within the past 10,000 years. Landslide hazard zones are defined either as (1) on or downhill from an active or potentially active landslide deposit, or (2) in geologic terrains that are known to be prone to landslides. These designations are not based on any specific recurrence interval or hazard intensity measure.

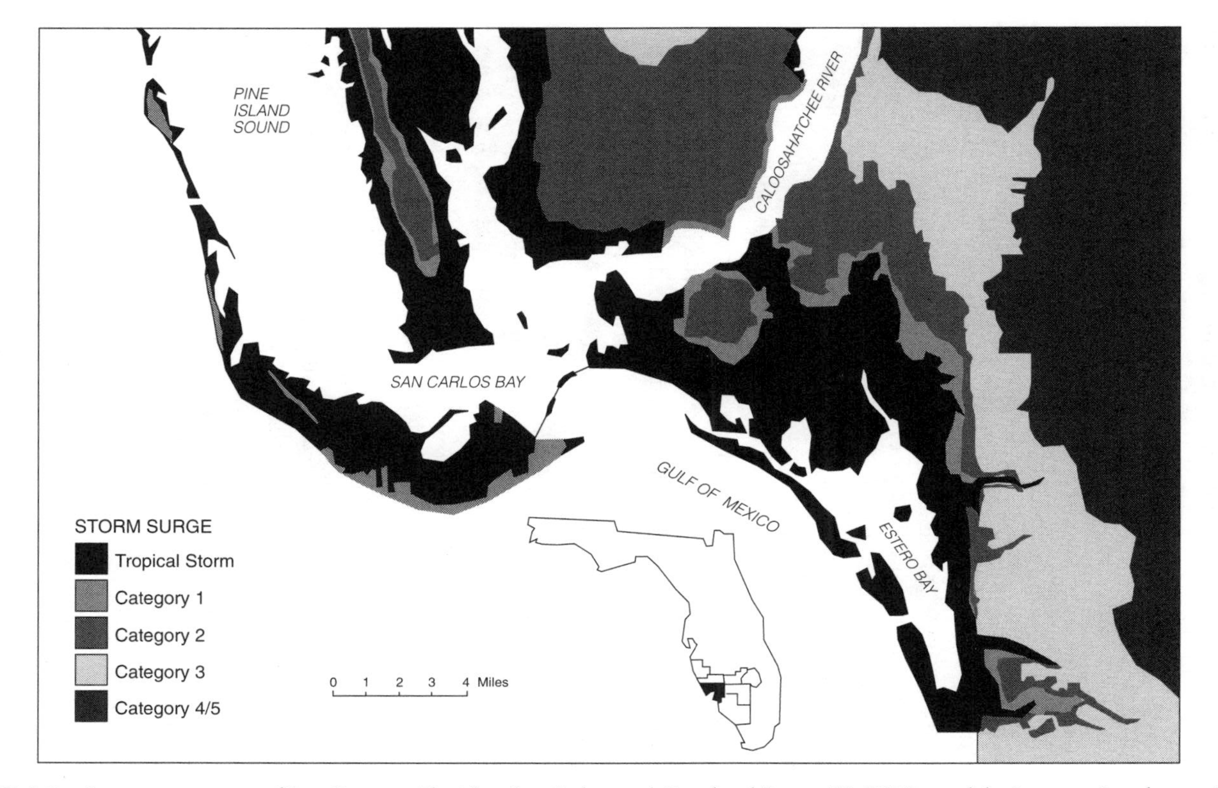

FIGURE 5-2 Storm surge map of Lee County, Florida. Sea, Lake, and Overland Surge (SLOSH) model. Source: Southwest Florida Regional Planning Council.

Simple hazard identification can be quite an effective tool for land use planning and management. Hazard identification is necessary for providing public information and for building commitment among elected officials. Awareness of hazards and their locations is the first step toward initiating land use solutions for the problems they pose. Hurricane storm surge maps defining evacuation zones are widely disseminated in hurricane-prone communities. Earthquake ground-shaking maps have been disseminated both in hard copy and on the Internet to inform the general public and local officials of earthquake hazard areas in northern California. Several states also use hazard maps to define the boundaries within which hazards must be disclosed. For example, California has an earthquake hazard disclosure requirement tied to real estate transactions in designated active fault zones. Massachusetts has proposed a similar requirement for property subject to coastal hazards.

Maps of hazards often are used in planning documents to demarcate areas within which specific land use policies are to be applied. For instance, Florida's 1985 growth management legislation requires communities to define coastal high-hazard areas within which they must analyze vulnerability and for which they are required to develop hazard mitigation goals, objectives, and policies as part of their comprehensive plans. The statute requires localities to base their delineation of the coastal high-hazard area on the area subject to flooding by storm surge associated with a Category 1 hurricane. The California Seismic Hazards Mapping Act requires that localities include seismic hazard maps in their general plans (Calif. Public Resources Code, chapter 2690).

Hazard maps also are used to define the area within which specific land use management tools are applied. Communities that participate in the National Flood Insurance Program prohibit development within the 100-year floodway and apply special building code provisions to structures built within the 100-year floodway fringe. Local governments in Florida are required to adopt and enforce certain minimum building code standards for flood-damage prevention and wind resistance within an area designated as the "coastal building zone." The zone is based on the 100-year velocity zone defined under the National Flood Insurance Program and, in areas with sandy shorelines, a setback from the area subject to erosion by a 100-year storm. Similarly, several coastal states define a zone subject to coastal erosion or flooding of a specified magnitude and probability as an area within which permits are required for land development or where other restrictions apply, such as setbacks from the mean high-water line. Model building codes include seismic hazard maps with zones that indicate the level of ground shaking that should be incorporated into equations for structural design. Portland (Oregon) is using seismic hazard maps to set priorities in implementing the state's new seismic-retrofit law.

Hazard identification also can be used to define areas in which more detailed studies are required prior to development. This approach defers decisions about mitigation from the scale of the community to the scale of individual building sites. California's 1971 Alquist-Priolo Special Studies Zone Act (renamed the Alquist-Priolo Earthquake Fault Zoning Act) defines special study zones, based on active fault zones, in which site-scale studies must be done. Within these zones, detailed geologic analysis is required as a part of the subdivision or building permit process for large developments. These special studies may then lead to the imposition of specific site-design or construction standards through a locality's zoning ordinance or subdivision regulations; at a minimum, the state prohibits new construction within 50 feet of an active fault. A similar approach has been followed by some communities for designated landslide areas (Olshansky, 1990).

Hazard maps can be used to identify property or structures to be acquired or relocated for hazard mitigation purposes. Hazard maps also are used to define areas within which federal, state, and local policies concerning critical and capital facilities apply. The federal Coastal Barrier Resources Act, and similar executive and legislative initiatives in Florida and Massachusetts, limit federal or state spending for growth-inducing infrastructure on hurricane-prone barrier islands. Florida's 1985 comprehensive planning mandate similarly requires local governments to limit investments in infrastructure in designated coastal high-hazard areas. In like fashion, where taxation initiatives have been directed at mitigating natural hazards, hazard maps have been used to define the areas subject to the policies.

Vulnerability Assessment

Vulnerability assessment, the second level of hazard assessment, combines the information from hazard identification with an inventory of the existing (or planned) property and population exposed to a hazard. It provides information on who and what are vulnerable to a natural hazard within the geographic areas defined by hazard identification; vulnerability assessment can also estimate damage and casualties that will result from various intensities of the hazard.

It is relatively easy to determine where people live, even though most residents may be at work, at school, or engaged in other activities for much of the day. Estimating the population of commercial areas or of tourist areas can be more difficult than estimating residential population, but reliable estimates that take account of daily and seasonal movements are generally possible. Over the past several years it has become clear that some segments of the population will be affected more than others by natural disasters (Bolin, 1993a; Tobin and Ollenburger, 1992). This presents significant equity issues that are not commonly accounted for in contemporary hazard assessments. To account for these differential societal impacts, it is important to conduct an inventory that captures the age, ethnicity, income, and relevant health characteristics of the population at risk. Recently a number of researchers have developed methods to estimate the societal impacts of natural hazards. For example, Perkins (1992) has developed a method for estimating the need for temporary housing in the event of an earthquake. French et al. (1996) have developed a method using small-area census data to assess the demographic characteristics of populations affected by damaged infrastructure.

When considering property at risk, it is generally desirable to consider the building inventory separately from the infrastructure systems that serve the community. Building inventories should include type and location of structure. Age also may be useful since it can often be related to the degree to which mitigation measures were incorporated (for example, in conformance with building codes in existence at the time of construction). With earthquake vulnerability assessment in particular, it is critical to classify the buildings by structural type (i.e., wood frame, steel frame, unreinforced masonry, etc.) because different classes of structures respond differently to given levels of ground-motion intensity. Structural classification is also useful when considering vulnerability to hurricane winds or wildfire hazards. Building occupancy may be important if life safety or casualties are to be considered; the use of the building is important if the assessment is to incorporate secondary economic impacts.

Infrastructure includes roads, bridges, water supply, sewerage, and electric-power systems. These systems generally consist of a variety of different components, each of which may exhibit a different response to a natural hazard. Only in recent years has it become possible to perform vulnerability assessments that take into account the performance of individual components of these systems (Patel, 1991; Sato and Shinozuka, 1991). Repair costs, replacement values, and length of service interruptions are important in estimating the economic vulnerability associated with infrastructure (Davis et al., 1982a, b; Seligson et al., 1990). Service-interruption estimates also are important in planning emergency response and in determining the demand for temporary food and shelter.

Vulnerability assessments attempt to predict how different types of property and population groups will be affected by a hazard. *Vulnerability functions* are empirically derived relationships that describe the response of populations, structures, or facilities to a range of hazard intensities. Examples of vulnerability functions include the damage-probability and restoration-time matrices for earthquake damage to buildings and infrastructure published by the Applied Technology Council (1985, 1991) (see Figure 5-3); the Standardized Earthquake Loss Estimation Methodology under development by the National Institute of Building Sciences, which can be used to estimate the probable number of deaths and injuries and potential economic and housing losses from earthquakes (National Institute of Building Sciences, 1994); and the flood damage tables developed by the National Flood Insurance Program (see Table 5-4).

One of the principal applications of vulnerability assessment has been the production of damage-loss assessments, which typically are used to estimate probable damages to private property from a natural hazard of one or several specified intensities under existing land use conditions. These are sometimes referred to as *deterministic* damage-loss assessments, which can be distinguished from *probabilistic* assessments, which assign probabilities to a whole range of likely events. Probabilistic assessments are the product of a full-scale risk analysis.

Deterministic damage-loss assessments have been done for hurricane-prone areas (e.g., Berke and Ruch, 1985; Ruch, 1983; U.S. Army Corps of Engineers, 1988, 1990; Withlacoochee Regional Planning Council, 1987) and for earthquake-prone areas (Association of Bay Area Governments, 1995; Central United States Earthquake Preparedness Project, 1990; Steinbrugge et al., 1987). They are primarily used to predict demand for emergency response services, public disaster assistance, and insurance losses. These loss assessments are usually done for hazards that affect large areas; they are not common for wildfires or landslides, for which more localized vulnerability assessments are typically conducted. Vulnerability assessments are also commonly used to assess structural alternatives for flood control within individual drainage basins (see, for example, South Florida Water Management District, 1984). Less common are studies that assess the impact of land use change on flood vulnerability (Linn, 1988).

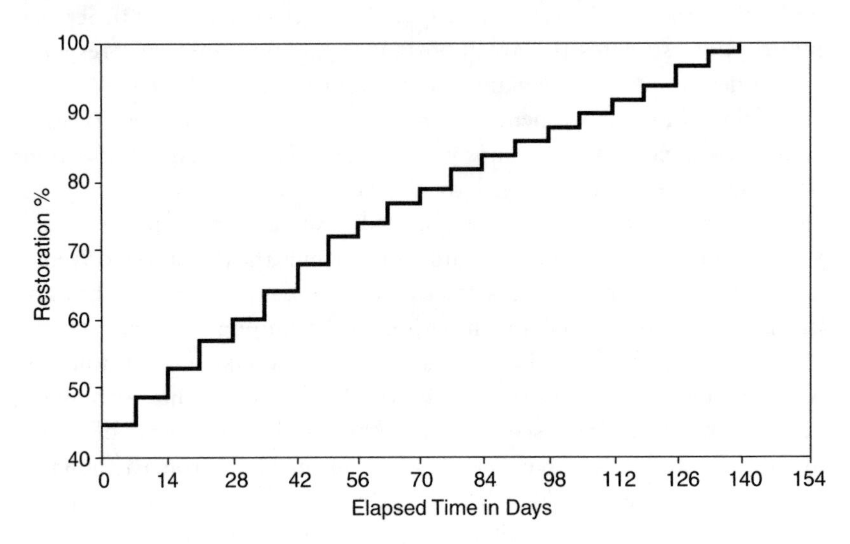

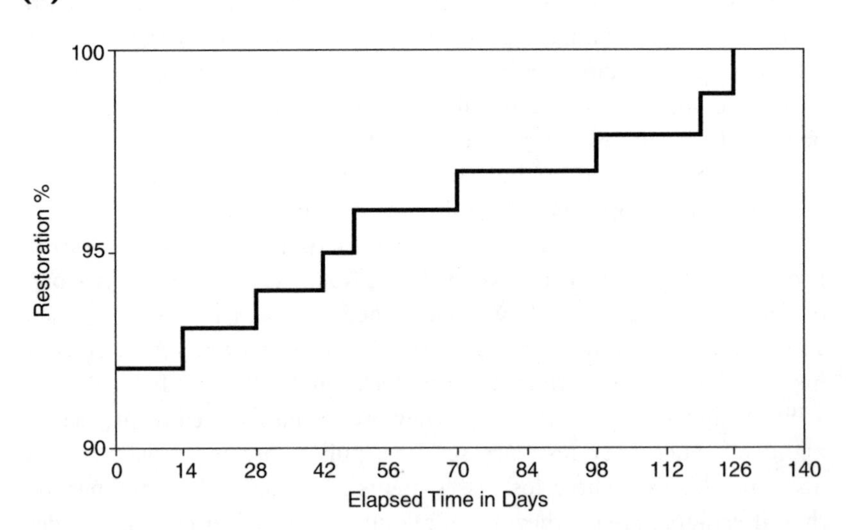

FIGURE 5-3 Restoration time for electric power following earthquakes of different magnitudes. (a) Residual capacity of Washington State's electric power following Puget Sound event (M = 7.5); (b) residual capacity of Missouri electric power following New Madrid event (M = 7.0) (Applied Technology Council, 1991, p. 386).

TABLE 5-4 Damage Percentages Based on Flood Depth (One-Floor Buildings, No Basement, Non-velocity Zones)

Water Depth (ft)	Percent Damage (based on claims data)
1	15.16
2	24.14
3	28.12
4	31.77
5	36.85
6	35.21
7	40.94
8	40.22
9	44.24
10	46.50
11	51.44
12	44.03
13	45.87
14	41.21
15	55.03
16	40.48
17	36.00
18	36.92

SOURCE: Federal Emergency Management Agency (1995a).

Vulnerability assessment can be applied across the full range of land use planning and management tools. It provides the fundamental data upon which emergency-response plans are based. For example, Florida's planning mandate links emergency planning and land use planning by requiring local governments to maintain or reduce hurricane evacuation times within designated coastal high-hazard areas. This requires detailed characterization of the populations within hazard zones, but not estimates of damage or injury. Forecasts of casualties and property damage for different land use scenarios may be used to design and justify a variety of public initiatives, for example: (1) building codes or other development regulations; (2) property acquisition programs; (3) policies concerning critical and public facilities; (4) taxation strategies for mitigating risks from natural hazards; and (5) information programs for members of the public who are at risk.

Risk Analysis

Risk analysis is the most sophisticated level of hazard assessment. It involves making quantitative estimates of the damage, injuries, and costs likely to be experienced within a specified geographic area over a specific period of time. Risk, therefore, has two measurable components: (1) the magnitude of the harm that may result (defined through vulnerability assessment); and (2) the likelihood or probability of the harm occurring in any particular location within any specified period of time (risk = magnitude × probability). A comprehensive risk analysis includes a full probability assessment of various levels of the hazard as well as probability assessments of impacts on structures and populations.

There are relatively few examples of probabilistic analyses of potential damages from natural hazards, although it appears this level of analysis is becoming more common. One example of such an application is the analysis performed by engineer Andrew Dzurik and his planning colleagues (1990) for Gasparilla Island in Lee County, Florida. Dzurik and his associates estimated the probable damage from each of five hurricane-intensity categories for a given year based on the value and structural characteristics of 461 existing habitable buildings and the probability of each storm category. They generated an annual probable loss value of $2.7 million based on 1986 land use. The Tampa Bay Regional Planning Council (1993) used a similar approach to calculate annualized losses from hurricanes for their region of Florida. Litan and colleagues (1992), in a report for the National Committee on Property Insurance, estimated the expected costs over 40 years of earthquake damages in Memphis and Los Angeles County.

Full-scale risk analysis can be used to support proposals for land use management policies or specific development projects, but there is little documentation of such use by local governments. Environmental impact analysis of proposed projects subject to the National Environmental Policy Act of 1969 (P.L. 91-190, sec.2; *U.S. Code*, vol. 42, sec. 4321), and similar state legislation, ideally should be based on assessment of the potential natural hazards at a site and the characteristics of the populations, property, and natural resources at risk. Benefit-cost analyses performed by the Army Corps of Engineers for flood control projects use probabilistic risk models to estimate damage avoided from floods of differing magnitudes. Federal regulations governing the siting of facilities such as nuclear power plants, hazardous waste management facilities, dams, and wastewater-treatment plants also require evaluation of the risks posed by floods or earthquakes.

Full-scale risk analysis also is appropriate for making decisions about hazard mitigation initiatives through property acquisition, design and siting of critical and public facilities, and taxation. Researchers with the Florida Planning Laboratory at Florida State University have worked with Lee County, Florida, to develop a risk-based fee system for county emergency management services (Smith and Deyle, 1994). This project used probabilistic models of storm surge and wind damage to personal property, public facilities, and infrastructure to devise a fee system linked to the risk incurred by development of land in different hurricane hazard zones within the county. Risk analysis also can be used to monitor the effectiveness of land use and other mitigation strategies over the long run. As development occurs, and as these programs are implemented, a community's expected losses are likely to change. Periodically conducting a full-scale risk analysis can provide decision makers with feedback on how their decisions are cumulatively affecting the overall sustainability of the community. While there are no extant examples of risk analysis being used in this manner, long-term monitoring of risk provides an interesting prospect as we look for measures of sustainability.

CHOICES IN APPLYING HAZARD ASSESSMENT

Planners, policy analysts, and other public officials must make a number of decisions in applying hazard assessment to land use planning and management. These include decisions about the precision of data and the geographic scale to be used in the analysis, recurrence intervals and the temporal perspective, and the level of hazard assessment to use. In some instances, these choices are made by federal or state agencies without careful consideration of the implications for local officials who use the analysis. In cases where analysis and decision making are conducted at the same level of government, decision makers and their staffs exercise direct control over these choices. In this section we discuss each of these choices and their implications in terms of utility to local planners and decision makers, political feasibility, and legal constraints.

Precision and Geographic Scale of Analysis

Choices about data precision and geographic scale for hazard assessment ideally ought to be a function of the uses to be made of the analysis. Public information and commitment building can be accomplished with relatively imprecise information about the characteristics of hazards and vulnerable populations and property. However, as a general rule, greater precision is required of the data as the size of the decision-making jurisdiction decreases. Thus, simplification of assumptions about the characteristics of hazards and vulnerable communities is more acceptable for state and federal policy analysis (for example, estimating costs of disaster assistance) than for hazard assessments done for local land use planning or management. At national, state, or regional scales it may be possible to use relatively crude definitions of hazardous areas and generalized data about populations and building stocks. Local land use planning and management, however, require greater levels of precision both to politically justify differential treatment of land in various locations and to meet the constitutional test of equal protection where land use is to be regulated.

The utility of hazard assessment may be compromised where the precision of data is inappropriate for users. For example, studies of hurricane losses are typically performed for multicounty regions (Ruch, 1983; U.S. Army Corps of Engineers, 1990). These studies use a number of simplifying assumptions about the relationships between broad categories of land use and the dollar value of property damage. Such assumptions are acceptable when providing broad estimates of potential hurricane losses, but they are of limited use to local governments for estimating losses within their jurisdictions under different land use or regulatory scenarios. As a result, there has been a decline in their production, at least in Florida (Michael McDonald, Florida Department of Community Affairs, Division of Emergency Management, personal communication, 1993). Scenario-based loss studies done for earthquakes have a similar need for precise data (Association of Bay Area Governments, 1995; Central United States Earthquake Preparedness Project, 1990; and Steinbrugge et al., 1987).

The precision and scale of hazard information are especially crucial to land use management initiatives. Olshansky et al. (1991), in a series of interviews with local planners, found that seismic-hazard maps must meet three criteria to effectively guide local policy decisions: (1) the level of hazard varies (some zones within a city or county are shown to be more hazardous than others); (2) this variation occurs regardless of the magnitude or location of an earthquake; and (3) each zone implies a list of feasible policy options.

Mapping precision—that is, the ability to define a line that separates areas with significantly different levels of hazard—is determined by the available data and the nature of the physical phenomenon that presents a hazard. Where hazard data are obtained from secondary sources, such as regional, state, or federal agencies, the density of data points may be insufficient for defining the precise boundaries of hazards. Greater precision may require expensive and time-consuming special studies. Thus, state and federal agencies should be particularly cognizant of the needs of local governments for precise data if they intend for hazard information to be useful in land use planning and management.

For some natural phenomena there is little spatial contrast in levels of hazards, thus it is difficult to define meaningful boundaries between hazardous and non-hazardous areas. For example, the intensities and probabilities of seismic ground shaking and hurricane winds do not vary significantly at geographic scales that are useful for land use planning or management. For flooding, intensity and probability vary over a spatial continuum that principally is a function of topography. Thus the differences in hazards across any particular zonal boundary—for example, the 100-year floodplain—can be relatively slight unless there is a dramatic topographic change. The spatial contrast for liquefaction and landslide hazards, however, is well defined because these hazards are a function of soils and geologic substrate. Similarly, wildfire hazard zones can be defined reasonably well on the basis of vegetative/fuel-load characteristics, orientation (e.g., southwestern slopes), and topography.

Where hazards do not vary spatially, it is difficult to justify drawing boundaries that determine allowable land use types or intensities or the imposition of different development regulations. Instead, building design is used to mitigate these hazards. Macrozones are defined to specify the forces to be used in structural design (see, for example, Figure 5-4).

Where the intensities and probabilities of hazards vary over space, as with flooding, planners and local officials must make more difficult choices about acceptable levels of risk. While the justification for regulating development within 100-year floodplains (with a 1 percent annual chance of flooding) versus 500-year floodplains (with a 0.2 percent annual chance of flooding) may be reasonably clear, the rationale for regulating property within the 100-year floodplain but not adjoining property in the 101-year floodplain is less so.

The intended use of the analysis and the desired level of precision determine the appropriate geographic scale of hazard assessment. A relatively coarse hazard mapping scale, such as 1 inch to the mile, is adequate for public information purposes and for fostering commitment to public policy initiatives. Such a scale also may be adequate for national or state policy analysis. Hazard information at a scale of 1 inch to 2,000 feet (the scale used for U.S. Geological Survey 7.5-minute quadrangles) usually is sufficient for general land use planning and for decisions about locating public facilities. However, at this scale it is difficult to distin-

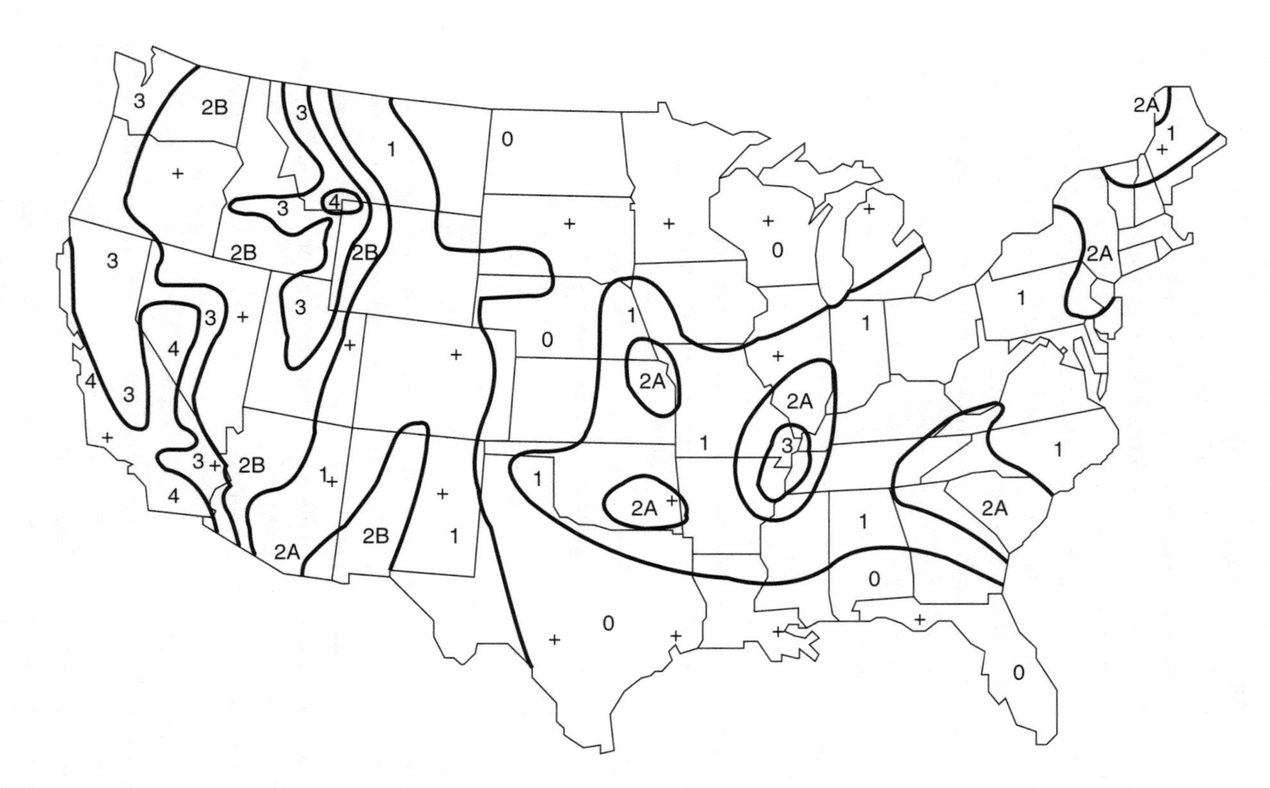

FIGURE 5-4 1994 Uniform Building Code zone map. Zones are identified by the numbers from 0 to 4. Seismic zone factors for structural engineering design are assigned to each zone. Source: Leyendecker et al., 1995, p. 27.

guish individual parcels of property. To support land use management applications at the level of the specific parcel, scales from 1 inch to 2,000 feet to 1 inch to 200 feet are required.

For planning and policy analysis by local governments, vulnerability assessment probably is done best using block-level census data and parcel-specific data on land use and structures. Census block data provide information not only on the number of people exposed at a level of resolution to fit hazard maps, but also some basic demographic facts on the age and ethnicity of the population. This information can be important in evaluating the social impacts of a disaster. Property inventories ideally ought to be done at the parcel level. More sophisticated analytic methods, such as geographic information system (GIS) technology, make it possible to use detailed levels of data available from secondary sources such as records maintained by tax assessors or inventories of public facilities. Substituting such detailed inventory data for generalized estimates should improve the precision of hazard assessments.

Unfortunately, increasing the precision of hazard identification and vulnerability assessment increases the cost of the assessment. Communities have to decide how much precision they are willing to pay for. Larger-scale hazard information can be used in land use management to identify areas where permit applicants are to be required to provide site-specific information. This strategy, however, forces local governments to depend on information provided by developers and their consultants. This may lead to problems with comparability among projects when different methods and standards are employed. Can a community support the types of planning and land use management it desires with general information that often comes from secondary sources? Do the benefits of more precise information justify the cost? The answers to these difficult questions depend on the severity of the local hazard and the community's commitment to deal with the problem.

Time Factors

Planners and analysts face two choices related to time factors in hazard assessment: (1) the recurrence interval of extreme events for which planning and development management should be done, and (2) the temporal perspective of the assessment, that is, whether to analyze risks involving current land use practices or future risks accompanying alternative land use scenarios.

All three levels of hazard assessment must take recurrence intervals into account. At one end of the temporal continuum are chronic hazards that produce modest levels of damage on a relatively frequent basis, generally with a recurrence interval of less than 20 years. At the other end are catastrophic events that recur less frequently, only once every several hundred years or more, but produce devastating levels of damage.

Choosing the appropriate recurrence level as the basis for planning and land use management is quite difficult for local officials. As a rule, their training and expertise is oriented towards short-term issues and decisions. The choice of a recurrence interval represents a value judgment that is difficult to deal with in the political arena. Many communities in California experienced difficulty in preparing the seismic safety elements of their general plans because they first had to select an acceptable level of risk. Communities do not face this problem in relation to flooding because of a federal mandate to plan for the 100-year flood, although they are free to select a more frequent recurrence interval. Most do so for the design of stormwater management systems, where facilities are typically designed for 5-, 10-, and 25-year events. Few communities design for floods that are less frequent than the 1-in-100-year event.

It is important to recognize that the recurrence interval is a function of the geographic scale from which one views the problem. While the 100-year flood only has a 1 percent probability of occurrence in any given year for a single community, at the national scale it is virtually certain that a series of 100-year events will occur each year. Thus, what appears to be a catastrophic event at the local level is a chronic problem at the state or national level. This difference in perspective lies at the heart of the disjuncture in hazard mitigation policy objectives that is the focus of Chapter 3.

The fact that recurrence intervals vary for different geographic areas also suggests that uniform land use policies ought not to be tied solely to hazard intensity. The contrast between seismic hazards in the central United States and California is illustrative. The New Madrid seismic zone can expect a magnitude 6.0 earthquake every 70 years, magnitude 7.6 every 254 years, and magnitude 8.3 every 550 years. The last magnitude 6.0 was in 1895. The last magnitude 8.0 was in 1812; earthquakes of magnitude 8.0 or greater are rare but can happen. The probabilities on the West Coast are much higher. For example, the Working Group on California Earthquake Probabilities estimated an 80 to 90 percent probability of an earthquake exceeding magnitude 7.0 in southern California before 2024 (USGS, 1995), and a previous Working Group estimated a 67 percent probability of an earthquake exceeding magnitude 7.0 in the San Francisco Bay region by 2020 (USGS, 1990). The City of Los Angeles can justify the expense of retrofitting all its unsafe buildings, because there is a high probability that it will save lives. In the central United States, the enormous expense and economic disruption are not warranted by the hazard. If we assume that buildings wear out in 50 to 100 years, then over time a policy of seismic design for new construction will eventually ensure that the building stock is seismic-resistant. This seems a prudent gamble in the central United States because of the time between earthquakes (an exception may be made for certain critical facilities), but it does not seem so for southern California.

The temporal perspective of hazard assessment as we define it here is not particularly relevant to simple hazard identification. It is, however, relevant to vulnerability assessment and risk analysis. The vast majority of vulnerability assessments and risk analyses are performed for existing land use patterns. In a few instances, researchers extend the analysis to predicting the impacts of forecasted growth according to existing plans. However, we have not found any examples where vulnerability assessment or risk analysis have been used to evaluate *alternative* land use scenarios. Dzurik and his associates (1990) estimate future annualized damages for Gasparilla Island (Florida) based on population growth projections. Berke and Ruch (1985) perform a similar analysis for Corpus Christi and Nueces County, Texas, based on a deterministic damage-loss assessment. Both groups of researchers suggest that their methodologies *could* be used to compare alternative future land use scenarios.

Level of Hazard Assessment

One of the important choices facing planners and policy analysts is the level of hazard assessment that should be conducted to support land use planning or management initiatives for hazardous areas. Hazard identification is the essential, minimum level of assessment needed to define the spatial extent of a hazard and to suggest any land use policies for mitigation of the hazard. It also is essential to generating commitment for such public policy initiatives. The product of hazard identification, which usually takes the form of a map, is relatively easy to understand, although as noted above, the utility of such maps depends on choices made about precision and geographic scale.

When and why should planners and local officials go beyond hazard identification? Vulnerability assessment ought to be used in the planning process itself because it is necessary to understand the consequences of alternative land use configurations. This type of analysis can measure damages that can be evaluated along with other aspects of each land use alternative in the development of a comprehensive plan, recovery or post-disaster reconstruction plan, or capital and public facilities policies. It also is important for setting priorities for property acquisition and for determining appropriate rates for impact fees or other revenue-generating policies linked to natural hazards.

Vulnerability assessment also is the minimum level of analysis required if a community desires to weigh the costs and benefits of various mitigation strategies. Hazard identification does not, by itself, provide the rationale for taking land use management initiatives. Vulnerability assessment, by providing information about the casualties and damages that may result from an extreme natural event in an area, provides the justification for deciding what level of hazard intensity warrants public intervention in private sector decision making. Vulnerability assessment is a stronger foundation for building commitment than is hazard identification because it provides decision makers with explicit measures of the potential impacts of natural hazards.

Vulnerability assessments require less information and less effort than full-scale risk analyses, and they are relatively easy to explain to decision makers, who usually are not well versed in probability theory. Vulnerability assessments, which typically analyze the losses from one or several hazard scenarios, do not formally incorporate the dimension of probability into hazard assessment. This does not mean that likelihood will be ignored, however. Instead, decision makers and their constituents will be guided by their perceptions of the likelihood of different hazard scenarios. Various studies have shown, however, that the perceptions of lay persons generally diverge fairly substantially from empirical probabilities (Sandman, 1985; Slovic et al., 1979).

Probabilistic risk analysis can provide a more accurate picture of the overall risk situation. By considering losses associated with the entire range of hazard intensities and probabilities, from the fairly common, low-intensity event to the extremely rare, catastrophic event, they capture the full range of potential casualty and damage experiences. Where a rational, comprehensive approach to decision making is desirable, such as through the use of benefit-cost analysis, the whole damage distribution should be considered based on a probabilistic risk analysis.

Aside from the technical difficulty and data demands of full-scale risk analyses, the principal constraint on their use is the ability of public officials and their constituents to comprehend them and apply them to policy choices. Berke and Ruch (1985) report that they purposely did not calculate annualized probable losses in conducting hurricane damage-loss assessments for several counties in Texas on the advice of state planners, who suggested that local officials would have difficulty understanding such information. Yet researchers such as Mushkatel and Nigg (1987a) have shown that the perceptions people have of risk, *and their behavior*, can be influenced by objective, quantitative risk information. Other researchers (e.g., Johnson et al., 1988) have found that fairly simple means of communication, such as booklets that present concise summaries of risks and mitigation alternatives, can change public perceptions and behavior.

Many hazard assessments are couched in such technical terms that they are not intelligible to local planners and engineers, much less elected and appointed officials or the general public. Hazard assessments must be presented in terms that are relevant to the policy choices faced by local officials (Panel on Earthquake Loss Estimation Methodology, 1989). This is a challenge that must be met by those who advocate land use planning and management as hazard mitigation strategies.

Of equal or greater importance, however, is the need for credibility. Public decision makers and their constituents must have confidence in the predictions and estimates derived from hazard assessment before they will be willing to make the difficult choices often required for hazard mitigation through land use management. Because risk analysis in particular is based on probabilities, it contains a significant amount of uncertainty. Many local decision makers are reluctant to impose restrictive land use controls given the political conflict often engendered by such controls and the degree of uncertainty present in damage models. They fear that their actions will not withstand political or legal challenges due to this uncertainty. Explicit consideration of uncertainty requires use of confidence intervals, which presents an additional challenge to risk communication.

Choices about the level of hazard assessment to use in land use planning or development management ought to be based on an understanding of the state of the art and the constraints that it may pose for understanding and mitigating different natural hazards. In the next two sections we discuss the major dimensions of the state of hazard assessment knowledge and information and the state of the art for specific natural hazards.

THE STATE OF KNOWLEDGE AND INFORMATION

Effective use of the different levels of hazard assessment in land use planning and management depends on the ability of scientists and planners to collect and analyze data for hazard identification, vulnerability assessment, and risk analysis. Significant advances have occurred in the past 20 years in scientific understanding of extreme natural events, availability of data, and capacity to model and analyze natural hazards, but gaps and uncertainties remain. When planning for and making choices about land use management in hazardous areas, these uncertainties must be recognized and communicated to those who make and are affected by land use policy decisions. Failure to acknowledge these uncertainties and limitations can result in inappropriate reliance on natural hazards assessment information. However, as noted in the preceding discussion of choosing the level of hazard assessment, awareness of these uncertainties and limitations also may impede the acceptance of more sophisticated levels of hazard assessment by elected officials and the public.

Knowledge about Extreme Natural Events

Scientists and planners identify hazardous areas and estimate the likelihood of extreme events based on their understanding of the fundamental causes and processes of natural phenomena. But scientific understanding of these phenomena is imperfect. What causes an earthquake? Where will the next one occur? Why do some years have more hurricanes than others? Which slopes will fail in the next intense rainstorm? What environmental conditions spawn firestorms? Scientists do not fully understand the causal processes of many of these natural phenomena. Or if the process is understood, the systems (dynamic atmosphere, heterogeneous earth) are too complex to be fully understood and modeled. In either case, lack of understanding of causation means that the ability to predict is limited too. Storms can be predicted several days before they occur, and floods and landslides can be predicted based on storm precipitation estimates, but we cannot predict where and when earthquakes or wildfires will occur with any precision.

Nevertheless, advances are being made in understanding the behavior of some natural hazard phenomena. Each new extreme event provides additional information that helps scientists understand underlying causal relationships. In addition, modeling initiatives in some fields are providing new tools for land use planning and management.

Data on Hazard Characteristics and Vulnerability

The lack of good causal models requires analysts to use statistical models of probability based on historic data to estimate the likelihood of natural hazards affecting specific geographic areas. Historic data are lacking, however, for many extreme natural events. Another impediment to hazard assessment is an inability to define hazardous areas at high levels of geographic resolution. Nonetheless, there has been a huge increase in information on hazard characteristics and vulnerable populations and property within the past 20 years, which has significantly advanced the ability to use hazard assessment in land use planning and management.

The dissemination of floodplain maps under the National Flood Insurance Program is the best known example of the new wealth of information available for hazard assessment. Substantial advances have been made for other natural hazards as well. The U.S. Geological Survey and the California Division of Mines and Geology have produced maps that rate the potential for earthquake ground shaking, liquefaction, and landslides for some localities in California, the Pacific Northwest, Utah, and other earthquake-prone areas. Wildfire hazard mapping has been done at the national level in Australia, at the state level in California, and at the local level in areas such as Boulder County (Colorado) and the East Bay Hills area in Oakland (California). The U.S. Army Corps of Engineers has assisted states and counties in developing storm surge hazard maps for areas exposed to hurricanes.

Increasingly the data needed for hazard assessment are available in digital form. This allows them to be manipulated with a geographic information system (GIS) or loaded into an automated damage-modeling system. Population and housing data are now available on CD-ROM at the block level from the U.S. Bureau of the Census (1991). In addition, many local governments have developed land parcel databases that include building and value information for all the properties within their jurisdictions, and many local public works departments have infrastructure inventory data on computer-aided design (CAD) systems or GIS.

The greatest problem with obtaining data on the characteristics of natural hazards is the necessity for extrapolating that data from a limited historic record. Thus, there is sometimes little data from which to estimate the probabilities of hazard occurrence and the magnitude and spatial extent of extreme natural events. Flood policy, for example, is based on the concept of the 100-year flood, but virtually no river in America has 100 years of record (and even more than 100 years of record would be necessary to make a statistically acceptable estimate). For some phenomena, researchers are able to identify prehistoric events, but there is no way of knowing whether all such events have been captured. For example, the recurrence of large earthquakes on the San Andreas Fault is well known, because the trace of the fault can be exposed and studied. In contrast, the number and size of prehistoric earthquakes on the Cascadia subduction zone off the coast of Oregon and Washington is virtually unknown.

For some natural phenomena the ability to accurately estimate the probability of extreme events of specific magnitude decreases as the geographic area diminishes in size. This may be the result of limited historic data for small geographic areas, as is the case for hurricanes at the county level or flooding in small watersheds. In the case of earthquakes, there is great uncertainty about where a fault will rupture and thus what the level of ground motion may be for any specific area.

For some natural phenomena, such as coastal erosion, riverine flooding, wildfire, and landslides and liquefaction associated with earthquakes, the magnitude and probability of the hazard vary greatly over small distances, so that regional values provide little useful information. For these hazards, detailed, local assessments are essential. In some areas, sufficiently detailed data exist for specific natural hazards so that accurate, small-area, hazard assessments can be conducted. Hazard data also may be inaccurate, however, where human actions have altered the magnitude or areal extent of the hazard, such as through increases in impervious surface area in small watersheds or changes in wildfire fuel loads on private lands.

Modeling Impacts

In addition to limitations on the availability of data on the characteristics of hazards, databases and models for vulnerability functions historically have been limited in quantity or quality. Furthermore, the data collection and processing tasks are enormous: one must gather statistics on damage, structural type, and exposure for thousands of damaged structures in each disaster. However, two relatively recent advances have made it easier for planners and engineers to conduct hazard assessments: (1) improved scientific understanding of vulnerability functions, and (2) the advent of geographic information system technology.

Through post-disaster empirical research and simulation modeling, significant advances have been made in understanding the vulnerability functions that describe the effects of floods, hurricanes, and earthquakes on people and the built environment. For other hazards, such as landslides and wildfire, there are no standard ways of describing intensity of damage. Damage-loss assessment models have been applied predominantly to private property (e.g., Berke and Ruch, 1985; King, 1991). While damage functions are available for major types of public infrastructure, only recently have researchers begun to develop models to predict the other public sector costs of response and recovery from natural disasters. The state of knowledge is markedly better than it was 20 years ago, but a significant amount of uncertainty remains in this area. For details about the state of vulnerability assessment, see the next section of this chapter.

The ability to use damage functions and other vulnerability functions to perform credible hazard assessments has been enhanced significantly by the development and widespread deployment of geographic information system technology. Traditionally building inventories were collected by a field survey of the hazard area. This was extremely expensive and time consuming. Modern GIS technology provides the ability to overlay a natural hazard area on a digital parcel map (French and Isaacson, 1984; Hwang and Lin, 1993). Typically the parcel layer includes information on ownership, value, and use of the structure. This allows the user to differentiate the population, structures, and infrastructure components exposed to varied levels of hazard. For example, using this technology it is fairly straightforward to determine with a high degree of precision the number, types, and value of structures that will be inundated by a 100-year flood event.

Most modern geographic information systems also provide the data management capabilities of a relational database. This allows the user to store and manage the large amounts of data required for a complete inventory of the people and property at risk. Using the complete inventory improves hazard assessment by eliminating the need to make estimates regarding the number of people and amount of property exposed to the hazard. This technology has become so accessible that virtually any planner or policy analyst can use it today, as long as a GIS database has been developed for the geographic area of concern. By the end of 1996, several GIS-based hazard assessment software systems were under development (see, for example, the descriptions of the HAZUS and RAMP systems in the section of this chapter on earthquakes, below.)

With the development of better damage models, improved availability of hazard information, and the advent of new technologies such as GIS and expert systems, staff capacity is becoming the limiting factor. Local governments routinely identify staff training among the most serious problems they encounter in implementing GIS and other advanced technologies (French and Wiggins, 1989). The staff expertise available to apply these techniques to natural hazards problems is likely to be a major constraint in many jurisdictions. Local governments will typically have three options to deal with this problem. First, they can add new staff who have skills and expertise that are readily applicable to hazard assessment. This likely will not be feasible for many jurisdictions with limited staff resources and demand for a wide range of planning expertise. Second, communities can invest in educating existing staff on the techniques of risk assessment. This is probably the most desirable course of action, except for the scarcity of means available to provide this education. Finally, communities can turn to specialized consultants who have the expertise. This is probably the most feasible solution, but it prevents the use of hazard assessment techniques on a routine basis. Thus, a combination of the second and third options will probably provide the best mix of expertise and availability.

ASSESSMENT OF SPECIFIC NATURAL HAZARDS

In this section, we summarize knowledge of the hazards posed by floods, earthquakes, landslides, hurricanes and coastal erosion, and wildfire, and the practice of hazard assessment for these phenomena.

Floods

Hazard assessment for floods begins with hydrologic modeling. The HEC-1 and HEC-2 models developed by the U.S. Army Corps of Engineers are the most widely used flood analysis models. They begin with the rainfall record for the local area and then simulate the likely discharge that will result from storms of various probabilities. This discharge is then routed through a cross section of the floodplain to estimate flood depths at various locations.

To do an accurate analysis of property at risk from flooding, the base elevation of the first floor of a structure is required. Where this is available in digital form, a GIS can be used to combine this information with data layers on the spatial location of the flood hazard and property data usually used for tax assessment purposes. There are, however, many local governments that do not have data on the elevation of buildings built prior to a community's entrance into the National Flood Insurance Program. Once the depth of flooding is known for buildings in an area, it is a relatively straightforward matter to calculate dollar damages using a vulnerability function (stage-damage curve) derived from past flood events. Stage-damage curves developed by the Corps of Engineers estimate damage as a percentage of value based on the depth of flooding experienced. Probable casualties, loss of life, and losses from business interruption are harder to estimate.

A key problem with traditional flood hazard mapping is that it is based on existing upstream conditions in the watershed. The amount of flood discharge depends on a number of factors including the duration and intensity of the storm event, the size of the watershed, and the amount of impervious surface area. The latter is largely a function of land use decisions. Thus, as upstream development occurs, the amount of run-off in the watershed increases; the depth, areal extent, and velocity of flooding increase in turn (see Figure 5-5). In small, urban watersheds, significant land use changes can dramatically alter flood discharge volumes. In such areas, frequent revisions to flood hazard maps are required. The Denver Urban Drainage and Flood Control District has tackled the watershed development problem by assuming complete development of each watershed for its models. As a result, its estimates of

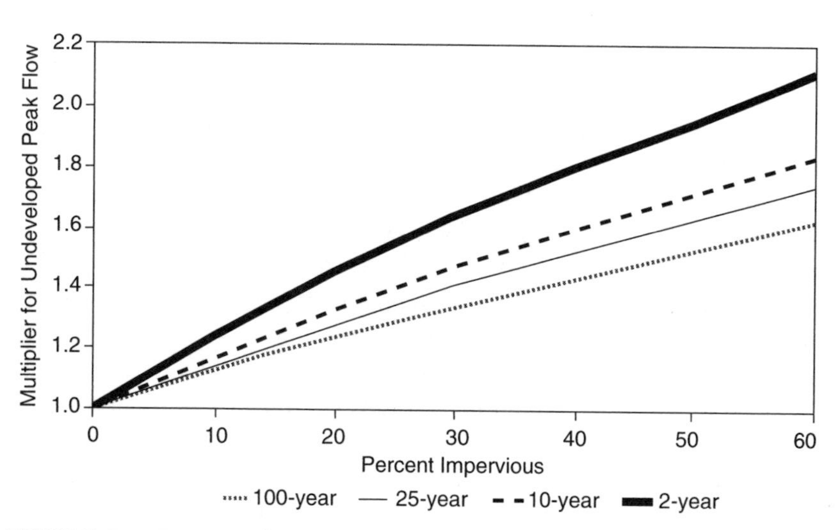

FIGURE 5-5 Impacts of urban development on peak flow. Source: Debo and Reese, 1995, p. 193.

flood depth and the geographic extent of its floodplains are greater than what is indicated on the Flood Insurance Rate Maps published by the National Flood Insurance Program. Conventional flood hazard mapping also does not account for the impacts from failure of flood protection works or the effects of debris in flood waters.

Earthquakes

Earthquake hazard assessment differs from other natural hazards in one important way: earthquake hazard is not as precisely tied to location. The hazard is diffuse, and wide areas share similar potential to be affected by an earthquake. As with floods, hurricanes, and wildfires, the structural characteristics of the property at risk are very important. Type and quality of construction play a major role in determining whether a building will be severely damaged by earthquake shaking. Thus, a building inventory is essential.

To perform an earthquake hazard assessment, seismologists and geologists must first identify earthquake source zones and assess long-term, magnitude-frequency relationships. This is complicated by the fact that some areas are exposed to earthquake hazards from multiple sources. In the central United States this is not a significant issue; there is a single major source, the New Madrid seismic zone. But on the West Coast analysts must account for the hazard posed by the San Andreas Fault system, large offshore subduction zones, and a multiplicity of smaller continental faults. Any given location has the potential to be shaken by an earthquake emanating from one of a large number of multiple sources. This makes it difficult to do a composite hazard map, although the Southern California Earthquake Center is doing just this. They have produced a map that adds up the hazard posed by all the faults and shows the probability of each location in southern California experiencing shaking greater than 0.2 g over a 30-year period.

The absolute size of an earthquake is expressed by a logarithmic magnitude scale based on seismometer readings. Because longer fault ruptures cause larger magnitude earthquakes, empirical formulas exist to estimate expected maximum magnitude as a function of observed fault length (Hanks and Kanamori, 1979). Given an earthquake magnitude, ground shaking at a site is estimated based on the site's underlying geologic material (Borcherdt, 1994) and its distance from the assumed earthquake source (e.g., Boore and Joyner, 1994). Locations farther away from the fault rupture shake less severely than locations closer to the rupture. Softer soil sites shake more severely than firm soil or bedrock sites. In general, damage and loss of life in recent major earthquakes have been concentrated in areas underlain by soft-soil deposits (Borcherdt, 1994). Estimates of ground shaking measured as effective peak acceleration are available at the county level, and models are being developed that predict ground shaking measured as peak velocity at the zip code and census tract levels (e.g., National Institute of Building Sciences, 1994). However, estimates of ground shaking are subject to great uncertainty at this scale, and as a result, they may be too crude for regulatory purposes.

Earthquake hazards generally can be grouped into three categories, all of which can be mapped: (1) direct damage from surface fault rupture; (2) damage caused by ground shaking; and (3) damage caused by shaking-induced ground failure. When fault movement occurs on the earth's surface, the structures built directly upon the fault trace will be damaged. Maps that identify active faults do not exist for most of the United States. California, however, has produced detailed maps of active faults, as required under the Alquist-Priolo Earthquake Fault Zoning Act (Hart, 1994).

Ground shaking is the most common cause of earthquake damage to buildings. From knowledge of past earthquakes, the Applied Technology Council and the U.S. Geological Survey have developed maps of expected forces of earthquake shaking throughout the United States (see Figure 5-6). Examples of these maps and descriptions of their historical development are given by Leyendecker et al. (1995). These maps form the basis for the seismic zonation maps used in all the seismic building codes in the United States. California, in 1991, enacted the Seismic Hazards Mapping Act, which calls on the state geologist to undertake a statewide seismic hazard mapping and technical advisory program (Real and Reichle, 1995; Tobin, 1991). This pioneering program in government-sponsored seismic hazard mapping is developing mapping guidelines and methods in several USGS quadrangles in southern California.

Strong ground shaking can trigger numerous types of ground failure, most notably landslides and liquefaction. These are some of the most hazardous effects of earthquakes, because they are not easily addressed by better design and construction of buildings. The U.S. Geological Survey has developed methods for generally assessing the susceptibility of slopes to earthquake-triggered landsliding (Keefer, 1984; Wilson and Keefer, 1985), but these maps only identify broad areas that could experience slope failures from a given earthquake. Liquefaction is a type of

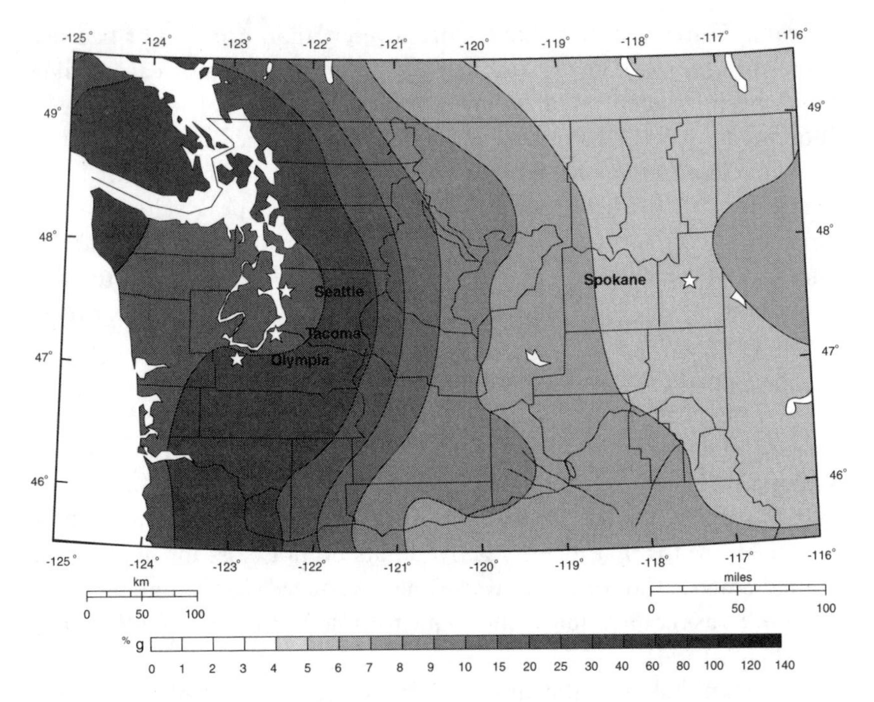

FIGURE 5-6 Map of Washington State showing the level of earthquake ground shaking that has a 10 percent chance of being exceeded in any 50-year period. It is shown in terms of peak horizontal acceleration, expressed as a percent of gravity. Source: Walter Hays, USGS, 1997.

ground failure in which soils temporarily lose their ability to support structures. Liquefied sands beneath the surface can cause differential settlement of overlying layers. Liquefaction also can cause lateral spreading, where blocks of overlying soil move apart and cause ground fissures. Because liquefaction occurs only in certain types of sandy soils, it is relatively easy to produce maps that show areas where liquefaction is possible when ground shaking exceeds a specified threshold. Such maps have become relatively common since the early 1980s, and many cities now include liquefaction potential maps in their comprehensive plans (e.g., Los Angeles County, 1990). One limitation of such maps is that they designate only how hazardous a zone is relative to other zones, rather than the expected extent of liquefaction that would occur in each zone in a given earthquake. Many areas within such a zone would not be susceptible to liquefaction. More detailed studies are needed to determine the presence of the hazard.

Earthquakes also can trigger secondary hazards, such as flooding and fires, caused by failures in structures. Maps showing the maximum extent of inundation from failed dams are common and are technically straightforward to prepare. The spread of fire after earthquakes, however, is not well understood, difficult to model, and typically not well addressed in earthquake loss estimates.

To assess vulnerability to earthquake damage, it is important to consider the type of buildings in the community. Low-rise, wood-frame buildings perform best, and unreinforced masonry (brick) buildings perform worst. Mobile homes also fare badly. Recently, reinforced-concrete frame buildings also have generated concern, as they can collapse catastrophically.

Under the National Earthquake Hazard Reduction Program, the Federal Emergency Management Agency, along with a consortium of professional organizations, has developed methods for evaluating the safety of existing buildings (Building Seismic Safety Council, 1992) and for designing new buildings for seismic safety (Building Seismic Safety Council, 1991).

Earthquake vulnerability assessments have been based on estimates of damage and injuries from single events. Until recently, these data collection and processing efforts were primarily for insurance purposes (e.g., Algermissen and Steinbrugge, 1978). However, damage estimates for given shaking levels have become more widely available since the early 1980s. For example, the Applied Technology Council's *ATC-13: Earthquake Damage Evaluation Data for California* (1985) provides detailed damage probability matrices that are much more advanced than what was previously available. (See also Association of Bay Area Governments, 1995; Central United States Earthquake Preparedness Project, 1990; Earthquake Engineering Research Institute, 1994; Steinbrugge and Algermissen, 1990; Steinbrugge et al., 1987.)

There are several problems, however, in using these assessments for public decision making. The methods of analysis vary considerably, so it is difficult to compare the assessments. They also are difficult and expensive to execute because each analysis must start from scratch. Both problems were addressed in a project for the Federal Emergency Management Agency led by the National Institute of Building Sciences (1994), entitled the Standardized Earthquake Loss Estimation Methodology. This project, with numerous consultants, oversight committees, and working groups, began in 1992, completed its work in 1997. The end result was a GIS-based, damage-estimation software system, known as HAZUS, designed to run on desktop personal computers, so that any city can perform its own loss estimate (*CUSEC Journal,* 1996). The user must provide the data on the geologic and built environment. The system is designed to run at various levels of precision, depending on the detail of input data, and it has default levels designed to give a first-cut approximation using generally available data. In addition to direct dollar losses, the NIBS system includes vulnerability functions for casualties, business interruption losses, secondary economic impacts, and social impacts such as the demand for sheltering. The system also can be used to perform formal risk analysis. A somewhat similar hazard assessment tool called RAMP (Regional Assessment of Mitigation) was under development in 1996 by EQE International, a consulting firm based in Irvine, California, under contract with the California Office of Emergency Services. RAMP is a GIS-based system that uses probabilistic models of seismic hazards and inventory data on critical facilities such as schools and hospitals to estimate annualized expected damage values. The designers indicate that the system will be capable of estimating the damage reduction benefits of alternative mitigation strategies and will be suitable for incorporation in benefit-cost analysis (Laurie Johnson, Urban Planner, EQE International, personal communication, 1996).

Landslides

Landslides occur in hilly regions and on river bluffs and sea cliffs all over the world. These slope movements pose a variety of hazards to human settlements. Fast-moving landslides are potentially deadly, because there is no time to warn occupants of downhill buildings. Slow-moving landslides can damage property but generally do not threaten lives. A single landslide does not affect as wide an area as floods or earthquakes. Patterns of landslide activity after a major storm or earthquake, however, can affect wide areas.

Landslide hazard assessment usually consists of hazard mapping only. These maps typically show areas with the potential for landsliding when the necessary threshold conditions of heavy rain or earthquake shaking are met. The assumption is that any structures affected by a landslide will be damaged severely. It is difficult to perform a quantitative vulnerability assessment or risk analysis because the uncertainty and imprecision of the hazard maps are too great.

Unlike floods, landslides do not occur in locations that are simple to define, and the same events do not predictably reoccur in the same places.

Frequency of landslide occurrences is usually unknown, and therefore predictability is poor. Some researchers have identified rainfall or ground-shaking threshold values that lead to good estimates of the frequency of events that can trigger landslides themselves at specific locations (Olshansky, 1990).

Although it is not yet possible to accurately estimate site-specific landslide probabilities, it is possible to identify hazard zones qualitatively. Brabb (1984a), Hansen (1984), and Varnes (1984) have published reviews of landslide hazard mapping methods. Most hazard mapping systems they report are based on the principle that if landslides have occurred in certain geologic environments, then all such environments in a region should be considered hazardous. In many cases, the hazard maps consist only of a landslide inventory, showing the areas that have actually had landslides in the past. It is still not possible, in the absence of costly subsurface investigation, to point to a given part of the landscape and identify the probability of landslide.

Landslide in Southern California (Walter Hays, USGS).

The U.S. Geological Survey, as well as several state geological surveys, have prepared landslide susceptibility maps in order to educate local planning agencies. The California Division of Mines and Geology prepared numerous landslide hazard maps under the Landslide Hazard Identification Program established by law in 1983. This program has now been incorporated into the Seismic Hazards Mapping Program.

Bernknopf et al. (1988) produced a landslide probability map for the Cincinnati area, which was a significant first attempt at producing a probabilistic landslide susceptibility map. This study also used damage estimates to evaluate the costs and benefits of identifying landslide hazards. The method used in this study probably could be replicated elsewhere but has not yet been repeated.

A few estimates have been made of the costs of landsliding, though all have been very approximate, and none has been based on a systematic inventory of people or properties at risk (e.g., Brabb, 1984b; Brabb and Harrold, 1989; and Schuster and Fleming, 1986). Petak and Atkisson (1982) systematically estimated landslide costs for the United States by combining a national landslide hazard map with county-scale population data. Probably the most systematic landslide hazard assessment to date was that done as part of the Urban Geology Master Plan for California, produced by the California Division of Mines and Geology (Alfors et al., 1973).

Hurricanes and Coastal Erosion

Hurricane hazard assessment has been hindered by the limited effort that has been made to develop data and models to support probabilistic damage models. By far the greatest effort has been exerted to support evacuation planning rather than hazard mitigation through land use and other controls. The principal hazard mitigation measures used have been building codes and, to a lesser extent, land use regulations such as setbacks; use of these measures has been based on deterministic models keyed to storms of a specified intensity or probability.

Hurricane evacuation planning is based principally on the delineation of storm surge hazard zones for each of the five categories of hurricane intensity (see Table 5-2). The Sea, Lake, and Overland Surge from Hurricane (SLOSH) model, developed by Chester Jelesnianski and Jye Chen of the National Weather Service, is used to calculate the maximum possible still-water flood depth resulting from the composite of an array of possible storms of a given intensity. Because SLOSH models are based on an elliptical telescoping grid method, the precision of estimates of surge heights varies from site to site. Jarvinen and Lawrence (1985) compared SLOSH predictions with actual surge heights in 10 storms at 523 sites and found that 79 percent of the errors were within 2 feet of the mean error of –0.3 foot.

Probabilities are not calculated for the different storms used to estimate these maximum-of-maximum storm surge inundation heights. Thus the hazard zones so defined represent worst-case conditions for a given storm intensity. No single storm of a given intensity category would inundate all the areas depicted on a SLOSH map, but at the time when evacuation decisions must be made, emergency management officials do not know precisely what the heading, forward speed, and landfall location of the storm will be.

It is widely acknowledged that storm surge inundation depths should be adjusted for wave height (Simpson and Riehl, 1980). Several rules-of-thumb have been used to estimate maximum likely wave height (Balsillie, 1983). Deterministic models have been developed to predict wave heights at specific locations based on near-shore bottom contours (Balsillie, 1983) and, for storm surge flooding, taking account of on-shore features such as dunes, structures, and vegetation (National Academy of Sciences, 1977). However, no attempts have been made to generate probabilistic estimates of wave heights at different locations for the full array of possible storms to which an area might be exposed. Storm surge inundation depths will also be affected by long-term changes such as sea-level rise and subsidence. Little research has been done to document these phenomena.

The state of the art for defining wind hazard zones lags considerably behind that for storm surge. While it is well known that hurricane wind velocities subside rapidly when hurricanes make landfall (Simpson and Riehl, 1980), only limited work has been done to model wind-speed degradation within the first 10 or 20 miles of landfall (e.g., Schwerdt et al., 1979). The most advanced wind degradation model for hurricane landfall (Kaplan and DeMaria, 1995) has been developed for land areas at the multistate, regional level and does not model wind-speed degradation at a spatial scale that is useful for hazard mitigation planning at the local level.

Coastal erosion hazard zones principally have been defined on the basis of long-term, chronic erosion rather than the erosion associated with individual storm events. Michigan, North Carolina, and South Carolina define regulatory zones for coastal construction based on some multiple of the average annual erosion rate determined by one of several methods of analyzing historic trends in erosion at specific coastal locations (Crowell et al., 1993). No probabilities have been formally estimated for these erosion-hazard zones, although the data probably would permit such estimates. In Florida, a regulatory zone for coastal construction is defined for sandy open shorelines by the area likely to be eroded by a 100-year storm. Location of the line is determined from analysis of historic data on storm and hurricane tides, maximum wave uprush, shoreline morphology, and existing upland development (Chiu and Dean, 1984). The Florida model could support development of a deterministic risk model.

Although the Federal Emergency Management Agency was directed to designate flood-related, coastal erosion–hazard zones under the 1973 Flood Disaster Protection Act, no such E-zones were ever designated. The original intent was to designate those areas subject to acute erosion caused by wave action. However, several bills that were introduced prior to enactment of the 1994 amendments to the National Flood Insurance Act recommended basing such zones on multiples of the annual erosion rate. In the 1994 amendments, Congress directed the agency to conduct a pilot project to evaluate the costs and benefits of defining erosion hazard zones.

The limited availability of data on the frequency of hurricane strikes makes hazard assessment difficult. The probability of a hurricane directly hitting any specific locality is very low (Jarrell et al., 1992), and the area directly affected is typically small; the radius of maximum winds around the hurricane center averages about 15 miles, and the direct hit zone averages about 50 miles (Hebert et al., 1993). As a result, estimates of the probabilities of hurricane strikes of specified magnitudes for geographic areas as small as counties or municipalities are based on Monte Carlo simulation using a model known as HURISK (Neumann, 1987; Science Applications International Corporation, 1989). These estimates are hampered to some extent by the limits of the historic database, especially for storm events with longer return frequencies than 100 years (Neumann, 1987, p. 53). However, the model output is readily accessible from the National Hurricane Center, and the model can be run on a PC by users who wish to perform their own analyses (Science Applications International Corporation, 1989).

Hurricane damage functions have been developed for still-water, storm surge flooding; velocity surge (waves); and wind associated with hurricanes. Most of these functions are based on broad categories of structures (e.g. one floor, no basement; two floors, with basement) rather than on specific structural characteristics. Most still-water-surge damage functions estimate structural damage from water height relative to the first finished floor of a structure (Federal Emergency Management Agency, 1995a ; Friedman, 1974, 1975; Friedman and Roy, 1966). The principal constraint on their application is the lack of data on first-floor elevations of structures within the geographic area of concern. Where property appraisal data are contained in a local government's GIS, first-floor elevations are likely to be included for structures built after a community began participating in the National Flood Insurance Program. Data for older structures are less likely to be available.

Wave damage functions also estimate damage based on first-floor elevations but consider both damage from flooding as well as the pounding effect of breaking waves (FEMA, 1981; Friedman and Roy, 1966). As with still-water-surge damage functions, the lack of comprehensive data on the elevation of structures makes it difficult to apply wave damage functions. Furthermore, as noted above, the data available are inadequate to define probabilistic wave hazard zones for the full range of possible storm events. (The V-zones designated under the National Flood Insurance Program represent only the area affected by a 100-year storm.) The Federal Insurance Administration wave damage functions, which are used to set V-zone flood insurance rates, are constrained by a limited database, the result of a relatively small number of claims for damage within V-zones (Howard Leiken, Chief Actuary, Federal Insurance Administration, FEMA, personal communication, 1996).

Richard Smith (1995) found substantial variation among the wind damage curves that have been developed, especially for predicted damage to mobile homes. Gary Hart (1976) offers the best documentation, but his curves are based on detailed information about structure type and building materials, most of which is not likely to be available in the databases of property appraisers. Damage curves developed by the U.S. Army Corps of Engineers (1990) are widely used.

Coastal erosion damage estimates are problematic because erosion is both a chronic and catastrophic phenomenon. The collapse of a structure from scouring and undermining of its foundation or pilings during a hurricane is the result of the progressive, long-term erosion of the site as well as erosion caused by a storm. Partly for this reason, and because most damage associated with hurricanes and other catastrophic coastal storms is due primarily to storm surge and wind, erosion damage curves have been difficult to develop. However, the Mobile and Philadelphia districts of the U.S. Army Corps of Engineers have developed such curves (U.S. Army Corps of Engineers, 1994). (For a comparison of the erosion damage curves developed by the two districts, see Coastal Planning and Engineering, 1992.)

The damage functions currently available for hurricane vulnerability assessments primarily apply to private structures and major categories of public infrastructure. Greenwood and Hatheway (1996) summarize a method developed by the Michael Baker Corporation for the Federal Emergency Management Agency's emergency management support teams, which was used during the approach of Hurricane Opal in 1995 to generate "real-time" maps and dollar estimates of damage to public and private structures based on damage functions. Similar approaches have been taken by Science Applications International Corporation (SAIC) in developing the Consequences Assessment Tool Set (CATS) for FEMA, and by Watson Technical Consulting in developing a hurricane damage model (TAOS) for the Organization of American States. SAIC's CATS model produces real-time and simulation estimates of ordinal levels of damage by structure type from wind storms and storm surge as well as earthquakes and various technological disasters (J. Pickus, Science Applications International Corporation, personal communication, 1997). Watson's TAOS uses a unique storm surge and wind model to estimate damages to various types of private structures and public facilities and is also capable of estimating debris generation and secondary economic impacts (Watson, 1995; C. C. Watson, Watson Technical Consulting, personal communication, 1997).

Several other initiatives are under way. FEMA is working with the National Institute of Building Sciences to extend their HAZUS earthquake loss estimation methodology to flooding and wind storms (see "Earthquakes," above). Researchers at the Florida Planning Lab at Florida State University are developing a model to predict the local costs of response and recovery to hurricanes for each of the public assistance categories under the federal Stafford Act (Boswell et al., 1997). Their model is based on key storm parameters and local land use and demographic information.

Wildfire

Wildfire hazards exist wherever development and wildland fuels meet. Over the last decade, wildfires have devastated over 20,000 square miles, consumed more than 6,500 structures, caused losses estimated in excess of $2.5 billion, and resulted in over 40 deaths. Yet wildfires in remote wilderness areas are of little consequence in many cases. In fact, from an environmental standpoint, these events actually may be beneficial: some ecosystems require periodic fire (for plant reproduction and nutrient cycling).

Wildfire experts have defined three types of wildfire hazard zones that land use planners should assess:

1. The *classic wildland-urban interface* exists where well-defined urban and suburban development presses up against open expanses of wildland areas. The 1985 Palm Coast fire (Florida) is an example of this hazard. The highly intense grassland-crown fire swept into a bordering community subdivision and destroyed 99 homes within just a few hours (Abt et al., 1987).
2. The *mixed wildland-urban interface* is more characteristic of the problems being created by exurban development: isolated homes, subdivisions, and small communities situated in predominantly wildland settings. The 1994 Chelan County (Washington) fire is an example, a fire which burned 2,970 square miles (nearly 8 percent of the county land area), destroyed 35 homes and countless nonresidential structures, and required mobilization of 10,000 firefighting personnel (Green, 1994).
3. The *occluded wildland-urban interface* exists where islands of wildland vegetation occur inside a largely urbanized area. Prior to the 1991 Oakland firestorm (which took 25 lives, destroyed 2,700 structures, and caused an estimated $1.68 billion in damages), the most notable example of an occluded wildland fire event was the 1933 Griffith Park fire, which took the lives of 25 firefighters in Los Angeles—the worst fire in the nation's history for firefighter casualties (Davis, 1988).

The lesson from the above typology is that very few communities are likely to be completely immune from wildfire hazards, because wildland fuels can be found in a variety of settings. Yet, as in the case of most natural hazards, hazard assessments for wildfires consist primarily of hazard mapping with some extension into vulnerability assessment. Moreover, wildfire hazard mapping itself is not widespread even in traditional wildland settings. To date, most mapping and simulations have been conducted to manage prescribed burns and plan response to fires, rather than for land use planning or policy making about managing development. Yet some notable mapping and mitigation efforts have specifically addressed wildland-urban-interface problems. These include the Australian Fire Hazard Mapping System (Morris and Barber, 1982), the California Division of Forestry State Responsibility Wildfire Hazard Mapping Program (Paul Barrette, Forester, California Department of Forestry and Fire Prevention, personal communication, 1996), the East Bay Hills Vegetative Management Consortium wildfire hazard mapping project (Radke, 1995), and the Boulder County (Colorado) Wildfire Hazard Information and Mitigation System program (Boulder County Land Use Department, 1994; Hay, 1994). These wildfire hazard mapping efforts vary considerably in scale, detail, and sophistication.

The Australian Fire Hazard Mapping System allows mapping of fire hazard to structures in mixed wildland-urban-interface areas on a map scale of one inch to 50,000 feet. The fire hazard model provides a crude form of risk analysis: estimates of the likelihood of fire occurrence are on an ordinal scale, and probable levels of damage are based on estimated fire intensity and suppression capability. Variables used in the model include frequency of fire season, length of fire season, slope aspect, slope steepness, vegetative cover and moisture levels, fire history, land uses (residential, forestry, etc.), egress from an area, and fire suppression capacity (Morris and Barber, 1982).

The California Division of Forestry has mapped most State Responsibility Areas (SRAs) of the state (about 35 million acres) for high wildfire hazard, taking into account topography, fuel loads, climate, and structural density (at scales of 1 inch to 1 or 2 miles) (Barrette, personal communication, 1996). Senate Bill 79 (1981) and Senate Bill 1916 (1982) require zoning of these predominantly unincorporated areas by county boards of supervisors according to the degree of fire hazard. The Division of Forestry also assisted numerous charter cities and counties in identifying 201 "very high hazard severity zones," as mandated by the Bates Bill, which was enacted after the 1991 Oakland firestorm in the hopes of preventing recurrence of such tragedies. The statute requires localities to identify very high hazard severity zones and adopt ordinances to mitigate wildfire hazards at a level of protection equivalent to or more restrictive than required by state law (e.g., vegetative clearance for a 30-foot perimeter around structures, spark arrestors on chimneys, and Class A roof materials). A subsequent California Attorney General opinion, however, emasculated the Bates Bill by interpreting the statute to require only identification of hazard areas, not official mapping of the hazard zones (Robert Irby, Deputy Chief, California Department of Forestry and Fire Prevention, personal communication, 1996).

The East Bay Hills Hazard Mapping Project is perhaps the most sophisticated occluded interface wildfire hazard identification project completed to date. Professor John Radke of the University of California developed two deterministic hazard models, one based on traditional wildland, fire hazard risk factors (e.g., topography, climate, weather, and fuel conditions) and the other based on expert judgments of urban/residential fire hazards (e.g., structural and vegetative fueling characteristics). Model outputs are integrated and embedded within a geographic information system to map regional and neighborhood vulnerability for existing and future development. The model also can be used to estimate the impact of mitigative actions on risk. According to Radke (1995), preliminary field analyses of the model have verified the accuracy of the hazard classification system.

The Boulder County Wildfire Hazard Information and Mitigation System (WHIMS) was created through the collaborative efforts of public and private entities to reduce wildfire risks to existing and future development. The WHIMS program uses a GIS-based, vulnerability assessment model to estimate the vulnerability of neighborhoods, structures, and lots to wildfire damage. County fire personnel conduct field surveys to generate much of the data needed for the hierarchical nested model. The model uses 27 variables to produce a vulnerability rating on a 0 (low) to 10 (extreme) scale (Hay, 1994). Model inputs include topographic traits, construction characteristics, vegetative cover, accessibility, and suppression capacity. Results from the model are used for fire prevention outreach to neighborhoods and individual property owners to identify effective mitigation measures to reduce vulnerability to acceptable levels (Boulder County Land Use Department, 1994).

In the future, planners may be able to integrate fire models into GIS-based wildfire hazard assessments. Fire models can describe many aspects of wildfire behavior useful for detailed vulnerability assessment and mitigation planning, including the rate of spread, future perimeter, area of the blaze, length-to-width ratio, and flame lengths (Fuller, 1991). Such modeling would allow land use planners to better understand the relative costs and benefits of alternative mitigation and development scenarios. However, existing models are only designed to consider wildland risk factors, not wildland-urban-interface factors (Finney, 1996; Fuller, 1991). Thus, high-risk meteorological and environmental conditions can be identified, but ignition points are impossible to estimate, and the list of potential sources is extensive (for example, lightning strikes, cigarettes, sparks from machinery, campfires, etc.). The relationship between wildfires and urban-interface factors such as structures and public infrastructure is not well enough understood at present to allow for integrated modeling (Radke, 1995). Yet, as the data themes necessary for modeling fires within the wildland-urban interface become increasingly easier to access in GIS-compatible formats, advances may be forthcoming.

IMPLICATIONS FOR POLICY FORMULATION AND IMPLEMENTATION

Land use planning is a complex process that requires the balancing of multiple objectives. For natural hazards to play a significant role in land use management decisions, the factual base detailing the nature and severity of the hazard must be at least as credible as that for the host of other issues that go into determining appropriate land use including real estate feasibility, economic development, transportation, service cost, environmental impact, and impacts on neighboring properties. Hazard assessment is the mechanism that provides this factual basis.

Clearly, a solid baseline of information that identifies hazards is needed for all hazards on a national basis. Regional, state, and federal entities can facilitate use of hazard assessment by providing much of this information, but it is clear that decisions about data precision and geographic scale will affect the utility of such information for local governments. The information will be most useful if it is at a scale of 1 inch to 2,000 feet or better. This level of information is currently available for floods and some aspects of hurricane hazards as a result of mapping carried out for over 20 years by the National Flood Insurance Program. While there is a substantial amount of earthquake information available, it is at varied scales and reflects a wide variety of recurrence intervals and different parameters for characterizing the ground motion. Landslide and wildfire information is available only for a limited number of areas. More extensive application of land use planning and management techniques thus requires a better factual base upon which to operate.

Local planners, policy analysts, and decision makers have important choices of their own to make about the data precision and geographic scale of hazard assessment, the temporal perspective and recurrence interval, and the level of hazard assessment to use. Where data available from secondary sources are not adequate, local officials must decide whether investment in better data is worthwhile. An important consideration in making such decisions will be the level of precision needed to politically justify, and legally defend, decisions about where to draw boundaries that determine allowable land uses or impose different development regulations.

An even more difficult decision concerns the appropriate recurrence interval that should be the basis for plan making and land use management. This choice crosses the line from technical issues to issues of policy as it directly relates to the question of acceptable risk. For flooding, national policy is to accept the 100-year (1 percent per year) standard; life and property need to be protected against any flood up to this size. The implication is that larger floods are too rare and costly to mitigate through land use means. For seismic hazards, building codes have been based on ground shaking with a 475-year return period. This is higher than for flooding, but it is a life safety standard only—buildings can be damaged, but not collapse. Acceptance of this standard implies that it is worth the expense to save lives for an earthquake up to this size, but that we cannot afford to design for larger earthquakes.

Regardless of whether local officials follow state or federal mandates or incentives, such as the National Flood Insurance Act, or choose recurrence intervals themselves, there is the risk that both public officials and citizens will treat the resulting boundaries, such as the 100-year flood zone, as absolute boundaries between hazardous and non-hazardous areas. This problem can be reduced where local officials and their constituents explicitly consider the trade-offs of alternative specifications of acceptable risk.

Making choices about acceptable risk, that is, what magnitude and likelihood of losses are deserving of public attention, raises a third issue: what level of hazard assessment should be used? One of the greater opportunities that has not been pursued extensively is the potential to use vulnerability assessment and risk analysis to analyze the risks associated with alternative land use plans. Similarly, these two levels of hazard assessment have the potential to provide better understanding of and commitment to land use planning and management initiatives to mitigate natural hazard risks.

Our review of the state of practice, summarized earlier in Table 5-1, reveals that current applications of hazard assessment in land use planning and management predominantly focus on hazard identification, and, to a lesser extent, on using deterministic vulnerability assessment to predict the impacts of extreme natural events of a limited range of intensities. Our review of the state of the art of hazard assessment shows that the last 20 years have seen significant advances in analytic capabilities and digitized land use data which make more advanced hazard assessment feasible. The principal constraints on greater use of these capabilities appear to include the following: (1) uneven knowledge of the probabilities, magnitudes, and locations of some types of extreme natural events; (2) limited parcel-specific data on relevant attributes of land uses, such as the type, design, and construction of buildings; (3) lack of empirically validated damage functions that are accurate at the building or infrastructure-component level for some natural hazards; (4) lack of professional expertise to incorporate sophisticated risk analysis models into land use decision making; and (5) lack of understanding and confidence in these models by appointed and elected officials. These findings suggest areas for research, education, and technical assistance initiatives by state and federal governments, educational institutions, and professional organizations.

CHAPTER SIX

Managing Land Use to Build Resilience

ROBERT B. OLSHANSKY AND JACK D. KARTEZ

LOCAL GOVERNMENTS IN THE United States always have had the lead role in influencing the process of community development. This is still true, even with the significant intervention into municipal and county land use regulation by a number of state governments since the late 1960s. The local role, however, has become increasingly complex.

Since the early 1970s, local governments have moved beyond zoning and subdivision regulations and traditional capital investments as their only mechanisms for guiding land use. Techniques and instruments have been added to the local implementation toolbox that can work in tandem with conventional regulation and budgeting to influence private development decisions. Today local governments make use of a wide array of additional regulatory, impact assessment, urban service control, and even educational tools to consciously guide the location, type, design, quality, timing, and distribution of costs of development. Taken together, the complex of tools and policies applied in each locality represents a *land use management guidance system* (Einsweiler, 1975). Twenty years ago little empirical work had been done on the use and effectiveness of land use management tools in mitigating losses from natural hazards (Baker and McPhee, 1975). In contrast, the intervening period has seen a large number of such studies, many of them focused on the effectiveness of these tools in mitigating hazards. This chapter presents a synthesis of that work.

In the following sections, we examine local use of a variety of land use management tools for hazard mitigation—their type, extent, and effectiveness, and factors that ease their adoption. Although the frequencies with which such tools have been adopted by local government can be easily summarized, it is much more difficult to reach conclusions about their effectiveness in mitigating hazards.

USE OF LAND USE MANAGEMENT TOOLS FOR HAZARD MITIGATION

Land use management has evolved to meet a variety of community needs, such as public safety, wise use of natural resources, and efficient public services. Nelson and Duncan (1995) identify 49 growth management techniques currently in use, with goals including resource preservation, special-area protection, rural growth management, urban containment, adequacy and timing of public facilities, and financing of public facilities. Hazard mitigation would seem to be another logical objective of integrated land use management.

Local governments, however, traditionally have not placed a high priority on hazard reduction. When local governments do address hazards, the impetus has often had to come from state and federal mandates and incentives, as discussed in Chapter 3. State mandates contribute to the quality of local plans, and can help to ensure that these plans address hazards, particularly when local commitment to act and the capacity to do so are present. But plans must still be implemented through specific mechanisms that are adopted and actually carried out locally, jurisdiction-by-jurisdiction. To carry out hazard mitigation plans, local governments must carefully choose an appropriate mix of land use management tools. Choices of the most effective implementation tools may not take place at all unless local government commitment and capacity are quite strong and problems are solvable. Local choices and the factors influencing them are profiled in this chapter.

Integrating Hazard Mitigation with Other Programs: An Example

An intentionally integrated set of land use management tools for hazard mitigation is still rare at the local government level. For example, although most local governments participate in the National Flood Insurance Program (NFIP), few have built the tools encouraged by the NFIP into an effective system of land use management procedures and regulations. One example of an integrated approach is the Tulsa, Oklahoma's floodplain and stormwater management program. By the 1980s, Tulsa was called the "flood capital of the world" because of repeated flooding of the same areas of the city. Another flood in 1984 that killed 14 and caused $180 million in damage finally galvanized local political interest in mitigation. As city official Ann Patton has described it, "The best time to stop a flood—or at least to cut your losses—is before the storm. That's why the City of Tulsa is doing its flood hazard mitigation planning now, before the water rises again. Flood hazard mitigation has many current shades of meaning. As used in the Tulsa program [it] is defined to mean [land] acquisition, relocation, floodproofing, and related actions taken before, during and after a flood. . . . [Tulsa] has found few model plans from other communities, although emerging federal policies tout the benefits of pre-disaster planning and nonstructural mitigation" (Patton, 1994). Tulsa's program pursues multiple objectives that contribute to overall management of development and redevelopment, such as neighborhood renewal, outdoor recreation/greenway development, and better utility service delivery. Tulsa's approach also is one of the small number of cases of local mitigation using housing relocation prior to the Midwest floods of 1993.

The Midwest floods of 1993 led the federal government to dramatically alter policy to support and fund permanent relocation on a large scale in a number of communities. Particularly noteworthy is that protection of the population (and taxpayers) from future flood losses was tied to the objective of preserving recreational and environmentally sensitive lands along the Mississippi River.

As discussed in Chapter 8, the emerging vision of environmental sustainability wedded to economic vitality and social equity is becoming an important rationale for reducing risk to human settlements from natural hazards. Tulsa's program, for example, has reduced future flood exposures while simultaneously improving recreation, neighborhood quality, and housing opportunity. That approach begins to realize the promise of sustainable development to meet human needs without squandering natural resources, damaging ecosystems, burdening future generations, or perpetuating injustices within society. The integrated use of appropriate land use management tools is necessary to achieve this vision, by reducing the intensity of development in hazardous areas, reducing the need for obstructing and altering natural processes, and reducing the social costs of vulnerable settlement patterns.

A Land Use Management Tool Kit

Local governments today use a wide variety of techniques to influence the location, type, intensity, design, quality, and timing of development. Many of these tools can help mitigate natural hazards. Local governments may choose from a range of approaches that can enhance a community's resilience, or ability to recover from hazards. These approaches include use of traditional zoning and subdivision powers to guide development away from hazardous areas, or to influence the type and density of development in hazard-prone areas. These traditional tools can also be used to regulate the design of sites and structures in hazardous areas. And many governments may choose to use some of the more sophisticated tools now available to encourage the private sector to plan and build more hazard-resistant developments. Hazard researchers have classified land use management tools into the following six categories:

- *Building standards* regulate the details of building construction. These typically are administered by building departments, although the standards are often triggered by requirements of planning departments. They include traditional building codes, floodproofing requirements, seismic design standards, and retrofit requirements for existing buildings.
- *Development regulations* are the traditional site development tools of current planning—the zoning and subdivision ordinances—which regulate the location, type, and intensity of new development. Development regulations can include flood-zone regulations; setbacks from faults, steep slopes, and coastal erosion areas; and zoning overlay zones which apply additional development standards for sensitive lands, such as wetlands, dunes, and hillsides. In some states, environmental impact assessment is used to assess site-specific hazards and recommend ways to mitigate their effects.
- *Critical and public facilities policies* affect public or quasi-public facilities, which local governments can control more readily than they can control private facilities. These policies include long-term capital improvement programs, siting of schools and fire sta-

tions, location of streets and storm sewers, and siting of public utilities. One obvious policy is to avoid placing public facilities in hazardous locations. In addition, because public facility siting is a key determinant of the location of new privately financed growth in a community, facilities should not be sited where they would encourage growth in hazardous areas.

- *Land and property acquisition* means purchasing properties in hazardous areas with public funds, and then using these properties in minimally vulnerable ways. This can include acquisition of undeveloped lands, acquisition of development rights, transfer of development rights to safer locations, relocation of buildings, and acquisition of damaged buildings.
- *Taxation and fiscal policies* are used to more equitably distribute the public costs of private development of hazardous property, specifically, to shift more of the cost burden directly onto the owners of such properties. These policies can include impact taxes to cover the public costs of hazardous area development, or tax breaks for reducing land use intensities in hazardous areas.
- *Information dissemination* is intended to influence public behavior, particularly, the behavior of real estate consumers. This category includes public information, education of construction professionals, hazard disclosure requirements for real estate sellers, and posting of warning signs in high-hazard areas.

There is a logic to the order of the above discussion of development management techniques: the more specific, traditional, and short-term strategies were described first, while longer-term, less site-specific policies were described last. The categories of development management techniques encompass a wide variety of approaches. Some emphasize long-range strategies, while others react to current development proposals. Some try to reduce development in hazardous areas, and others accept such development but focus on site design as a solution. Some redirect public investment, but most seek to regulate or influence private development. Some are regulatory, others are voluntary. Where hazards can be clearly delineated (see Chapter 5), the most appropriate land use management tools are those that reduce land development in high-hazard areas (such as setback regulations, public facilities location policies, and acquisition and/or relocation of hazardous properties). The remainder of this chapter evaluates the use and effectiveness of these tools for hazard mitigation.

Research Findings

Table 6-1 presents the proportion of local governments in the United States that have reported use, since 1975, of specific land use management tools to mitigate hazards. The table draws from several studies of hazard-prone areas.

Tables 6-1a and 6-1b, on pages 176–178, describe in more detail the data on which Table 6-1 is based, drawing on several major cross-sectional survey studies conducted between 1979 and 1993. These were studies of local government approaches to floodplain hazards (Burby and French, 1981; Burby and French et al., 1985), coastal storms and hurricanes (Godschalk et al., 1989), earthquakes (Berke and Beatley, 1992a) and across multiple natural hazards (Dalton and Burby, 1994; and Kartez and Faupel, 1995). The sample frames, types of specific informants, and survey response rates differ among the studies (see Table 6-1b). But all involved cities and counties as the units of analysis, used sample selection methods that allow some statistical generalization, and had very high survey response rates of 75 to 90 percent, permitting confidence that results are representative of the sample population in each study. Table 6-1b identifies characteristics of the sample frames in each study that are important to interpreting the results.

This summary of findings from a number of studies reveals some important patterns. Since 1975 the number of different hazard mitigation tools used by local governments has increased. As pointed out in Chapter 2, zoning was the basic tool in use for decades before the National Flood Insurance Program spurred the widespread extension of local land use controls to floodplain management. Table 6-1 suggests how local approaches have expanded from regulation alone to occasional use of strategies such as public facilities location and public and developer-builder education efforts as part of land use management for hazard mitigation. Still, the most frequently used tools are regulation of private development and construction in hazard areas.

Some land use management tools appear to be used infrequently or hardly at all except in the states vulnerable to coastal hurricanes. A climate has arisen in some communities that is more conducive to aggressive modification of development through a mixture of controls and incentives (as in Nags Head, North Carolina; see Chapter 4). Factors contributing to development of such a climate include the high economic returns from coastal development and high growth rates, more accurate hazard zone delineation, and state programs and state and federal man-

TABLE 6-1 Application of Development Management Tools to Mitigation of Natural Hazards

Development Management Tools	Percentage[a] of Localities Reporting Use of Each Tool, for Communities Subject to		
	Floods	Hurricanes	Earthquakes
Building Standards			
Floodproofing requirements, or special seismic building code	+++	—	++++
Development Regulations			
Zoning ordinance	++++	++++	++++
Subdivision ordinance	++++	++++	++++
Shoreline setback regulations	—	++++	—
Special hazard area ordinance, flood hazard area ordinance or fault-setback ordinance	—	+++	++
Dune protection regulations	+	+++	—
Wetlands protection regulation	++++	—	—
Natural hazards identified in city environmental assessments	—	—	—
Critical and Public Facilities Policies			
Capital improvements program	—	++++	+++
Location of public facilities outside of hazard areas (to discourage development)	++	+++	++
Location of public facilities to reduce damage risks	—	+++	+++
Land and Property Acquisition			
Acquisition of open space/recreation/ undeveloped lands for mitigation	—	+	+
Relocation of existing hazard area development	+	+	+
Acquisition of development rights or scenic easements	—	++	+
Transfer of development potential from hazard areas to safer sites	++	++	++
Taxation and Fiscal Policies			
Reduced or below-market taxation for open space or reduced land use intensity in hazardous areas of sites	+	++	+
Impact taxes or special assessments to fund the added public costs of hazard area development	—	+	+

continued

TABLE 6-1 *Continued*

Development Management Tools	Percentage[a] of Localities Reporting Use of Each Tool, for Communities Subject to		
	Floods	Hurricanes	Earthquakes
Information Dissemination			
Public information program	+++	—	+++
Construction practice seminars or builder/developer mitigation education	—	++	—
Hazard disclosure required in real estate transactions	—	+++	++

[a]++++, Most communities (greater than 50 percent); +++, many communities (25 to 50 percent); ++, some communities (5 to 25 percent); +, few communities (less than 5 percent); —, no data.

SOURCES: Berke and Beatley, 1992b; Burby and French et al., 1985; Godschalk et al., 1989.

dates for coastal zone planning. By contrast, using acquisition and relocation to change land usage in hazard-prone areas occurred very rarely (e.g., in Tulsa) until the great Midwest floods of 1993. In earthquake-risk areas as well, aggressive modification of new development patterns has been rare, because hazard zones cannot be so clearly delineated (see Chapter 5).

Other tools also are used only infrequently because of their costs—both monetary and in terms of the political or organizational efforts required to carry them out, rather than because they challenge property rights. These include land and property acquisition and, in particular, taxation and fiscal policies. Burby et al. (1991) studied local government approaches to the problem of paying for hazard damages to public facilities that were serving hazard area development. These researchers found little local interest in adopting hazard zone impact taxes or fees. Local officials viewed levying such taxes as much less feasible than purchasing insurance for future public facility losses, including self-insurance through loss reserves. Yet less than 50 percent of localities had actually purchased flood insurance for public facilities, and less than 25 percent had insured for earthquakes as of 1988. Very recently, Lee County, Florida, proposed an emergency public shelter impact fee to be based on area differences in risk to life and property and the consequent demand placed on emergency public shelters. As an impact fee, it is designed as a one-time fee for funding capital facilities only (Lee County Division of Public Safety, 1993). Lee County, whose public shelter impact fee is one of the first of its kind, has a history of interest and initiative in allocating the costs of public services in proportion to use throughout its growth management system. Applying impact fees to hazards should be seen as evidence of the county's commitment to and expertise in growth management generally, in the context of a very strong statewide planning law in Florida.

Seismic planning experts George Mader and Martha Tyler evaluated the state of earthquake mitigation through planning and land use management for a congressional review of the National Earthquake Hazards Reduction Program in 1991. They observed that even in California "the local use of [land use management] strategies is largely hit or miss without state mandates" (1993, p. 99). Mader and Tyler identified 11 of the more frequently used land use management and planning tools that they judged to be central to improving local efforts in California. Techniques such as reduced taxation and impact taxes were not included and indeed have not seen much use at all anywhere.

But Mader and Tyler pointed out that many existing land use management tools that are widely used at the local level could contribute more to hazard mitigation if each tool was simply extended to include a hazards component. An example is the use of environmental impact assessment (EIA), which all cities and counties in California must consider when taking any action subject to the California Environmental Quality Act (CEQA) (Calif. Public Resources Code, section 21000). In California, EIAs almost always consider seismic hazards. Kartez and Faupel's 1995 survey of city hazards management programs shows that EIA review procedures by local governments in California consider hazard mitigation objectives 95 percent of the time. Reviewing the environmental impact assessment can lead local officials to require specification of mitigation measures before projects are approved. Other states and localities mandate EIA reviews, but outside California only 44 percent of planning departments conducting these reviews report considering hazards in the process.

Mader and Tyler argue that "important lessons for local governments elsewhere are to be found in California's experience." One of these lessons is the recognition that existing and routinely used land use management tools can be used for hazard mitigation. "A key advantage to these strategies is that they require only relatively minor adjustments to normal local government practice to be effective. Once adopted, they

TABLE 6-1a Application of Land Use Management Tools in Mitigation of Natural Hazards, 1979–1993 (Estimates from Multiple Studies)

Development Management Tool	Percent of Localities Reporting Use of Each Tool (Study Authors/Data Year/Hazard)						
	Burby et al./ 1979, 1983[a]/ FLOOD		Godschalk et al./ 1984[b]/ HURRICANE & FLOOD	Berke & Beatley/ 1986[c]/ SEISMIC		Kartez & Faupel/ 1993[d]/ ALL	Burby et al./ 1991[e]/ ALL
	1979	1983		CA	U.S.		
Building Standards							
Floodproofing requirements, or special seismic building codes	59	41	*	69	0	*	83
Development Regulations							
Zoning ordinances	75	58	88	56	66	*	68
Subdivision ordinances	75	51	86	52	59	*	*
Shoreline setback regulations	*	*	54	*	*	47	35[e]
Special hazard area ordinance, flood hazard area ordinance, or fault-setback ordinance	*	*	27	21	8	39	35[e]
Dune protection regulations	6	3	38	*	*	5	*
Wetlands protection regulations	25	13	*	*	*	38	*
Natural hazards identified in city environmental assessments	*	*	*	*	*	44	41
Critical and Public Facilities							
Capital improvements program	*	*	54	27	33	*	*
Location of public facilities outside of hazard areas (to discourage development)	22	8	31	15	10	39	*
Location of public facilities to reduce damage risks	*	*	46	26	9	73	*

Land and Property Acquisition							
Acquisition of open space/recreation land/undeveloped land for mitigation	34	17	29	2	3	20	*
Acquisition of damaged buildings in hazard areas	*	*	3	0.00	1	*	*
Relocation of existing hazard area development	3	2	2	2	1	*	*
Acquisition of development rights or scenic easements	*	*	14	5	2	18	*
Transfer of development potential from hazard areas to safer sites	6	6	21	15	4	*	3
Taxation and Fiscal Policies							
Reduced or below-market taxation for open space or reduced land use intensity in hazardous areas of sites	9	3	11	0.00	5	*	2
Impact taxes or special assessments to fund the added public costs of hazard area development	*	*	2	0.00	2	*	*
Information Dissemination							
Public Information program	37	32	*	37	13	22	41
Construction practice seminars or builder and developer mitigation education	*	*	15	*	*	12	*
Hazard disclosure required in real estate transactions	*	*	26	17	6	40	9

*No comparable data on the tool in this study.

[a]1979 survey by Burby and associates includes only local governments with populations of 5,000 or more; hence, the median population is six times that of 1983 sample (17,592 vs. 3,221) (1979 data from Burby et al., 1981; 1983 data from Burby and French et al., 1985).

[b]Sample frame focuses on coastal states, which have high storm hazard incidence (Godschalk et al., 1989).

[c]Data are reported separately for sampled communities in California versus for communities in the rest of the U.S. (Berke and Beatley, 1992a).

[d]Sample frame limited to municipalities with populations of 20,000 or greater (Kartez and Faupel, 1995).

[e]All forms of hazard setbacks included (Burby et al., 1991).

TABLE 6-1b Characteristics of Local Hazards Management Surveys

Study and Hazard Focus	Sample Frame	Response Rate	N	Type of Informants
1. Burby and French et al., 1985 (Flood)	In 1979, 1,415 cities and counties	85%	1,203	Local flood coordinators
2. Burby and French et al., 1985 (Flood)	In 1983, 1,219 cities and counties	78%	959	Local food coordinators
3. Godschalk et al., 1989 (Hurricane)	In 1984, 602 localities	67%	403	Local planning director or designee
4. Berke and Beatley, 1992b (Earthquakes)	In 1986, 104 communities in California; 156 communities in other states	83%,, 78%	85, 122	Local planning director or designee
5. Kartez and Faupel, 1995 (All Hazards)	In 1993, 375 municipalities surveyed in 1986	90%	337	City planning director or designee
6. Burby and May et al., 1997	In 1991, 176 counties in California, Washington, North Carolina, Texas, Florida. Half (88) in planning mandate states and half (88) in non-mandate states	98%	176	Local planning director or designee

readily become institutionalized, making consideration of seismic hazards a routine matter in approving new development or building modifications" (Mader and Tyler, 1993, p. 98). That advice applies to land use management strategies for other types of natural hazards as well.

EASING ADOPTION OF LAND USE MANAGEMENT TOOLS

For a variety of reasons, mainly perceptual and political, local governments resist taking action to regulate today's private property for the promise of a safer tomorrow. It is possible in many areas to identify land vulnerable to natural hazards, and many techniques exist to restrict land uses in these areas. Identifying vulnerable areas and restricting the use of those areas would be considered rational behavior. But, human society does not necessarily behave rationally. Natural hazards are not high on the political agenda because they occur infrequently and are overshadowed by more immediate, visible issues. Seismic safety or flood hazard mitigation have rarely elected a governor or mayor.

Nevertheless, mitigation policies *do* get adopted by local governments. Many communities *have* implemented them successfully. What can we learn from these successes? What are the ingredients for success? What factors help communities to enact and enforce land use restrictions in hazardous areas? How can citizens move their local governments to take action?

Over the past 20 years a number of researchers have attempted to identify the key factors that influence the adoption of hazard mitigation policies (Alesch and Petak, 1986; Berke, 1989; Berke and Beatley, 1992a; Burby and French et al., 1985; Dalton and Burby, 1994; Drabek et al., 1983; Godschalk et al., 1989; Hutton et al., 1979; Lambright, 1984; Olshansky, 1994; and Wyner and Mann, 1986). Their results are as summarized in this section.

Berke and Beatley (1992a) classified these key influencing factors into two groups, depending on whether they are under the control of local government. The controllable factors include: public recognition of the problem; persistent, skillful, and credible advocates; repeated interaction and communication among participants; availability of staff resources; and linkage to well-established precedents, or linkage of hazard policy issues to conventional ones. Uncontrollable factors include: occurrence of a disaster that leads to a "window of opportunity" for change; community wealth and resources to support new mitigation initiatives; political culture that supports regulation of private property for

The New Madrid earthquake of 1811 was actually a series of 200 moderate to large earthquakes which occurred from December 1811 to March 1812. Although these quakes caused little loss of life because the region was sparsely settled, the area affected by the New Madrid quakes was two to three times that of the 1964 Alaska earthquake, and ten times as large as that affected by the 1906 San Francisco earthquake (State Historical Society of Missouri, Columbia).

the public good; mandates or assistance from state or federal governments; previous experience with hazards; and presence of a policy solution that is both technically and politically feasible.

Lambright (1984) fitted some of these factors into a chronological sequence of policy development. In a study of seismic policy environments in southern California, South Carolina, and Japan, Lambright outlined the course of earthquake policy innovation as follows: (1) an awareness of a problem by a few "entrepeneur" individuals; (2) a trigger that converts awareness into policy demand; (3) a search for an appropriate response; (4) the adoption of a policy option by government, thereby giving legitimacy to the policy; (5) the translation of policy into program action; and (6) the institutionalization of programs.

All of the studies mentioned agreed that the key initiating factors are the presence of an advocate to promote awareness, and a window of opportunity to trigger action. However, it should be noted that Burby and May et al. (1997), in an empirical study of communities in five different states, found that previous occurrence of a natural disaster did not have a strong effect on the number of mitigation techniques employed by communities.

Most of the cited studies address hazard mitigation in general, and most involve just one type of hazard, usually earthquakes or flooding. The lessons theses studies derived may be extended to all hazards, and the factors that influence all forms of mitigation can be applied to the narrower category of land use regulation.

Factors Controllable by Local Government

Political and institutional roadblocks to hazard mitigation are formidable. Studies of earthquake hazard mitigation have suggested various reasons for the lack of action, even where concern is high: earthquakes themselves are not well understood, there is no public constituency, costs are immediate and benefits uncertain, benefits may not occur during the tenure of current elected officials, public safety is not visible, and other public issues are more immediate. But these roadblocks can be overcome.

Recognition of the Problem

Problem recognition consists of awareness and concern, and clearly both are necessary prerequisites to local action. Local governments and community organizations, with advocacy and staff resources, can persuade members of the public and elected officials to recognize the importance of natural hazards. But awareness and concern alone are not enough. May (1991) summarizes the results of seven surveys of earthquake awareness and concern in the western and central United States and concludes that all consistently demonstrate that earthquakes are expected but not dreaded. Officials and the public alike are aware that they face an earthquake risk, but it is not a high-priority concern.

Study findings vary regarding the level of concern of local officials. Drabek and others (1983) found that most community leaders sampled in Missouri and Washington felt that seismic risk is a serious problem, and most were concerned that an earthquake could result in major damage and loss of life. Wyner and Mann, in a more limited study (1986), found that most elected public officials and city managers in California considered seismic risk to be moderate or high. Mushkatel and Nigg (1987a, b), in a 1983 survey, found that a substantial proportion of citizens and government officials in the New Madrid seismic area were at least somewhat worried about the possibility of a damaging earthquake before the year 2000. They also found high levels of support for mitigation measures, with strongest support in the areas most at risk. In contrast, Rossi and others (1982) documented that natural hazards have a very low priority on the local political agenda, and most officials have little interest in land use approaches to mitigation.

Even where local officials state that they are concerned, they may not be concerned enough to take action. For example, although Drabek and his colleagues (1983) found high levels of concern for earthquake hazards in Missouri and Washington, the level of earthquake mitigation activity among state agencies was quite low in both states.

Some studies have contrasted the level of concern among officials with that of the public. Mushkatel and Nigg (1987a) found a higher level of concern for seismic safety among residents than among public officials. Ironically, Drabek and colleagues (1983) and Wyner and Mann (1986) found that government officials *perceive* their awareness and concern to be higher than that of the citizenry.

Governments can improve community-wide recognition of hazards by preparing local comprehensive plans, as described in Chapter 5. Burby and May et al. (1997) found that high-quality plans result in greater knowledge and more understanding of natural hazards among stakeholders in the community. And the more knowledgeable they become, the more likely they are to act on that knowledge.

Advocates

Numerous studies have emphasized the importance of key individuals, acting as advocates or policy champions. Even in their relatively pessimistic assessment, Rossi and others (1982) observed that the viewpoints of key influential people are important in advancing natural hazards on the local agenda. Alesch and Petak (1986, p. 225), in their insightful study of earthquake mitigation policy in Los Angeles and Long Beach, concluded, "The probability that hazard mitigations will be enacted is in direct proportion to the extent that there are inside policy advocates who are persistent and tenacious in their pursuit of the policy, who have access to policy makers, and who have credibility among policy makers."

Interaction Among Participants in Policy Development

Policy development depends on many actors, including agency personnel, professional associations, and civic organizations. Sustained interaction and communication among participants is important in helping to define issues, develop solutions, and bring them to the attention of others. Interaction can consist of repeated communication between agencies, or exchange of information through activities in professional associations. It can take place through organized coordinating bodies, such as seismic advisory councils, as highlighted in the following chapter.

Staff Resources

Limited resources hamper mitigation efforts. Both money and staff time are needed to adequately develop and implement local mitigation programs, even when data and assistance come from external sources. When resources are available, data can be collected and analyzed, solutions developed, and mitigation actions carried out. It is important to be able to retain well-trained local staff, who can develop, analyze, and present the data, and who are familiar with local problems. Local officials need to be committed to mitigation programs and need to have adequate capacity to implement them.

Linkage to Other Issues

Mitigation of natural hazards is more likely to succeed if it can be seen as reinforcing the solution of another problem, such as power plant siting, school safety (which has been important regarding seismic safety efforts in Arkansas, Kentucky, and Missouri), growth management, or traditional techniques of local land use management. Hazard mitigation programs also are eased by the existence of well-established precedents. Once a precedent is established for one field, accustoming people to new ways of doing things, it can promote similar change in another field.

Factors Not Controllable by Local Government

Many factors that affect the use of hazard mitigation measures are beyond the control of local government. Unlike the controllable factors noted above, which have been explored via detailed qualitative case studies of local processes, these other factors are more easily measured by quantitative empirical means. In many ways, they are easier to measure and explain, but less relevant to local governments seeking actions they can take to facilitate mitigation policy.

Burby and French (1981), in a nationwide study of flood-prone communities, concluded that those that have adopted broader, more stringent programs are communities where flood-prone land is currently in use. Also these communities perceive the flood risk as serious, attach high priority to solving flood problems, have taken other steps to deal with flooding, and have a relatively greater capacity to mount a vigorous program. Many of these factors are inherent in the existing qualities of a community: larger, wealthier communities with vacant upland areas are more likely to have substantial floodplain management programs. Paradoxically, they also found that communities with the strongest development pressures were stimulated to adopt strong floodplain land use management programs.

Godschalk et al. (1989) measured three types of factors affecting mitigation of coastal flooding: environmental factors, policy catalyst factors, and political conversion factors. Their analysis identified three environmental factors that negatively influence mitigation (percentage of floodplain already developed, presence of barrier islands, and existence of a strong private property ethos). With regard to the policy catalysts, their analysis identified five positive factors (recent storm history affecting priority and adoption of mitigation actions, probability of a hurricane, community advancement to the regular phase of the National Flood Insurance Program, number of years in the NFIP, and state active in promoting mitigation) and one negative factor (recent storm damage causing planners to rate their programs as ineffective). The authors identified two negative political conversion factors (absence of politically supportive groups, and opposition of development interests).

Opening of a Window of Opportunity

Numerous hazard policy studies have emphasized the catalyzing effect of the window of opportunity that opens following a natural disaster. Disasters increase local awareness and concern and attract federal and state resources. Sometimes distant disasters can create windows of opportunity, as a community is reminded of what may someday occur closer to home. Proactive local agencies can plan for mitigation programs in advance, so that they can be introduced into the policy arena shortly after a window opens (and before it shuts again). California's Seismic Hazard Mapping Act, for example, was outlined in the late 1980s, making it ready for rapid introduction and enactment following the 1989 Loma Prieta earthquake. As noted earlier in this chapter, Burby and May et al. (1997) found counter-evidence, to the effect that previous occurrence of a natural disaster did not have a strong effect on the number of mitigation techniques employed by communities. Still, they recognize the value of other types of windows, related more to internal than external factors: plan updates, comprehensive revision of zoning ordinances, or major capital investment decisions.

Community Wealth and Resources

Community wealth can influence local mitigation initiatives in three ways. First, in affluent communities, the costs of mitigation may be seen as acceptable compared to the benefits of loss reduction. Second, a prosperous community with a strong tax base can afford to initiate new, innovative programs. Third, local officials with a strong tax base are willing to restrict development in hazard areas even though it may shift some development to other jurisdictions.

Political Culture Supporting Regulation of Private Property for Public Ends

Communities that are less "public minded," more individualistic, and more concerned with protection of property rights tend to give little support to hazard mitigation. This is particularly true of mitigation programs that regulate land use type and intensity. The difficult political and legal problems associated with private property rights considerations are discussed in more detail below. In general, the success of local hazard mitigation plans requires that citizens and elected officials become committed to the planning process and feel a sense of ownership of the plan.

Mandates or Assistance from State or Federal Government

Natural hazards are generally of more concern to higher levels of government than to local governments. The wider the jurisdiction of the government, the more likely the area is to experience disasters; for smaller areas disasters are low-probability events. The federal government must respond to disasters every year, whereas a given local government rarely experiences such an event. Thus, it is in the interests of state and federal governments to encourage local mitigation programs. There have been numerous studies on the effect of federal policies that facilitate local mitigation programs. In some cases, federal programs provide funding or technical assistance. In other instances, the state or federal programs mandate local regulatory and public investment actions. According to Burby and Dalton (1994), local land use plans, to be effective vehicles for mitigation, usually require state or federal mandates, as discussed in detail in other chapters of this book.

Previous Hazard Experience

Communities with previous disaster experience are more likely to be aware and concerned enough to take mitigation actions. This is particularly evident in coastal (hurricane-prone) regions, although the influence of strong state planning and hazard-related mandates in those states (Florida, North Carolina) makes it difficult to ascribe local action to experience alone. Kartez and Faupel (1995) found evidence in a longitudinal study of 400 cities in 1987 and 1993 that land use management actions to mitigate hazards take five to ten years to become institutionalized once a major disaster has caused damages. Noteworthy cases like Tulsa illustrate that multiple, costly impacts are often necessary for experience to foster action.

Presence of a Feasible Policy Solution

The final ingredient for success is the existence of a policy solution that is feasible, both technically and politically. Alesch and Petak put it as follows: "In order for hazard mitigation policy to be enacted, there must be an available policy option that includes a technical solution viewed as practical and efficacious by nontechnical policy makers" (1986, p. 224).

Land Use Regulation: Special Considerations

Although only some of the land use management tools identified in this chapter involve applying land use regulation to private property, this type of regulation is potentially the most powerful means of mitigation, because it reduces the number of people and properties in risky locations. But it also is problematic, both legally and politically.

Legally, local governments need to worry about claims that they have taken the value of private property, and hence must compensate the owner. Although the courts have not formulated a clear test that distinguishes between a land use restriction so severe that it constitutes a taking and a permissible restriction upon the use of private property, courts have generally allowed land use regulations that substantially advance legitimate public interests, do not deny owners economically viable use of their land, and do not unduly burden individuals. Thus, to avoid being subject to claims of taking, hazard-related land use regulations should clearly serve a legitimate public interest, be supported by scientific data demonstrating a connection between the regulation and the public interest, and should not render land valueless. In addition to meeting legal requirements, it is often politically prudent to maintain economic value of property to the maximum extent possible consistent with the public safety goals of the regulation (this subject of private property rights is discussed in Chapter 2; also see Kusler, 1980).

THE QUESTION OF EFFECTIVENESS: LOCAL IMPLEMENTATION, AS DISTINCT FROM ADOPTION

Most of the discussion to this point has looked at factors that lead to policy adoption, and not necessarily successful implementation. May and Bolton (1986) argue that planners need to think of implementation when considering the likely effectiveness of hazard reduction measures. With regard to earthquake hazard reduction measures, they assert that theoretical studies "have overstated the benefits of such measures. By assuming away implementation difficulties, the studies provide a 'best case' analysis of the theoretical effectiveness of hazard-reduction measures" (p. 450). For example, studies after Hurricane Andrew found that the South Florida Building Code was adequate, but actual construction workmanship and code enforcement were deficient (Federal Emergency Management Agency, 1992a). Focusing on adoption alone, therefore, may be unrealistically optimistic. For example, Wyner and Mann (1986), regarding local implementation of seismic mandates in California, found that very few jurisdictions had taken steps to achieve even some of their seismic land use objectives. Palm (1981) found that fault-zone regulations in California, although successful in limiting new construction of homes directly upon active faults, did not affect consumers; they affected neither housing prices nor consumer decisions to buy houses in these zones.

What, then, are the ingredients of successful implementation? Failure in the implementation of local plans is not an issue unique to hazards, of course, even when state mandates exist to force local action. May and Bolton (1986) draw on the implementation literature to warn planners that complex chains of implementing actions and indirect control can undermine implementation, but conclude that the difficulties can be surmounted if the program creates incentives for target groups and key implementers. As an example, they concluded that a proposed hazardous building abatement ordinance in Provo (Utah), though superficially attractive, would be unlikely to work in reality, because key target groups probably would not comply.

A few studies have identified specific pitfalls in trying to regulate land uses for hazard mitigation. For example, Wyner and Mann (1986) concluded that implementation of seismic land use policies does not actually stand in the way of efforts to develop land. Communities prefer building code or site-specific design approaches to land use approaches. Local governments favor regulations that pass costs on to developers on a project-by-project basis. As noted in Chapter 5, regarding local use of spatial hazard information, planners are most likely to use information that (1) is clearly mapped, (2) comes from an authoritative source, and (3) provides guidelines for specific actions they can take.

Burby and others (1988) found that floodplain management programs are effective in protecting *new development* from flood losses *up to the 100-year event*. Some lessons from the successful programs: (1) benefits are achieved primarily through influence on the development decisions of builders and land developers—once land is subdivided, habitation and use will follow; (2) the post-flood window of opportunity is best used for targeting households and business firms to retrofit or relocate; (3) NFIP criteria do not discourage floodplain development, but local land use programs often exceed the requirements of the NFIP in discouraging floodplain development; (4) floodplain development pressures can be reduced if cities ensure that a large supply of flood-free land is available for urban development; and (5) land acquisition is always an option.

It is difficult to generalize from these few studies of implementation, but some tentative lessons from what actually works are listed in the following paragraphs.

Use clear and authoritative maps of the hazard

Maps should be clear and unambiguous, so that planners and public officials can tell which zones apply to a particular property. A credible scientific body or expert, who is seen as being unbiased, should issue the maps. Examples include: flood hazard maps issued by the Federal Insurance Administration (although local governments with more specific engineering data may modify the maps); fault-zone maps issued by the California State Geologist; and informational landslide maps prepared by the U.S. Geological Survey and several state geological surveys.

If trying to revise or restrict land uses in hazardous areas, do so before the land is subdivided

Once land is subdivided, the individual parcels may be sold and owners are entitled to the use of their property. If some of the parcels are wholly within hazardous areas, the only way to prevent their development is for the local government to purchase the parcels. Some local governments, under pressure from developers, allow subdivision of properties with the condition that subsequent applications will be carefully reviewed on a parcel-by-parcel basis. If detailed study were to show that one parcel, for example, is located at the foot of a large active landslide, the problem would be unfair and costly to city and property owner alike.

Use project-specific design approaches

Although well-designed comprehensive plans can establish the policy context, the most feasible approaches are actually those that can be implemented on a project-by-project basis. These approaches are easy for local governments to administer because they can be integrated with normal development review processes, and they do not involve a large number of other institutions. The types of policies that are most likely to be implemented successfully by local governments are policies that call for clustering of development on the least hazardous parts of a property, design of subdivisions sensitive to natural processes, building setbacks, site-specific engineering studies, and building elevation or strengthening. One problem with project-by-project implementation policies is that, like dams and levees, they may encourage development of hazardous areas by creating a false sense of complete safety and thus increase the potential for catastrophic losses. Engineered safety is only provided up to a specified design standard (see Chapter 5, regarding acceptable risk).

Link clear and realistic design guidelines to the map

Local planners need to know what to do with a hazard map. The clearer the instructions, the more likely they will be to follow them. The more realistic the guidelines, the more likely they will be accepted by the community. Floodplain ordinances, for example, usually have detailed lists of design requirements for each zone, such as building elevation, prohibition of storage of hazardous materials, and further studies needed.

Offer incentives to encourage target groups to comply

Incentives for developers might include tax abatements, density bonuses, or waiver of off-street parking requirements. At the very least, programs, in order to be successfully implemented, must be designed to be sensitive to the needs of target groups.

Ensure that hazard-free land is available for development

If an entire jurisdiction is in a flood zone or a zone of high seismic hazard, then it probably would not be feasible to apply strict development restrictions over all properties. Cities need safety valves for growth pressures, and every jurisdiction with hazard areas should also have relatively safe areas designated for development.

Use the post-disaster window of opportunity to encourage individual owners to retrofit or relocate

Individuals are most aware of the hazard in the immediate aftermath of disaster. Owners of tornado-damaged homes are more likely to consider wind-resistant design, for example. The strong desire on the part of many floodplain communities to be relocated following the 1993 Midwest floods is another good example of this phenomenon.

If hazardous land is subdivided and built out, be prepared to purchase selected properties

Property acquisition is always an option. If individual parcels are found to be in highly flood-prone areas, coastal erosion zones, or actively unstable slopes, public agencies should be prepared to purchase the properties. This often is more cost-effective than waiting for extensive property damage, injuries, or loss of life to occur, all of which may also be accompanied by costly litigation.

The above principles are the result of extensive assessments of experience and indicate what *ought* to be the approach to implementation under some typical conditions.

CONCEPTUAL AND OBSERVED EFFECTIVENESS

To create a report card of the comparative effectiveness of different land use management tools is difficult for several reasons. First, the types of land use management techniques that are part of a local jurisdiction's tool kit differ both in their potential or theoretical effectiveness and their observed effectiveness. In theory, some land use management tools would seem to be more effective than others if consistently and correctly applied. For example, acquisition of hazardous lands would seem to provide a more direct solution than public information systems.

More importantly, it is hard to rate each tool, because the tools are not independent of each other. They can work together, mutually reinforcing each other's contribution, for example, a zoning control and a coordinated building code prohibiting development on unstable slopes (unless the design meets geotechnically safe design requirements).

Other techniques, like educating builders and investors about safe siting and design needs, would not seem a priori to be very effective alone in spurring private hazard mitigation, yet may work well when combined with other measures. Empirical data indicate that education used in concert with regulation is likely to be more effective than regulation alone, as Burby and Paterson found in a detailed study of the effectiveness of construction run-off and sediment control regulatory programs (1993). In fact, not using the education tool may cause regulation, which theoretically speaking ought to be successful, to be ineffective in actuality. Regulation can fail when the regulated community is uncooperative or ignorant of requirements, and when the public is politically unaware or unsupportive of the purposes of regulation. As part of a system, education is an essential and effective land use management tool; alone, education is a weak tool. The reality of land use guidance in the United States is that no system can rely on regulatory control alone. Burby and May et al. (1997) found that good land use plans help to educate the public, and that this knowledge may lead both to increased public demands for action and to increased commitment of planning staff and elected officials to the planned policies. Education is thus crucial from planning through implementation, and regulations adopted in jurisdictions with uninformed and uncommitted officials and citizens are unlikely to succeed.

The relative costs of the tools are also important. The most effective tools may simply be too expensive to provide widespread safety. For example, tools that remove hazardous lands from development, such as land acquisition or purchase of development rights, are too expensive to be used for all high-hazard areas.

Another reason why the observed (empirical) effectiveness of land use management techniques is difficult to summarize is that their effect depends on context. A good example of how effectiveness is influenced by context comes from the careful investigations of Burby and his colleagues into the effectiveness of local government floodplain management carried out within the framework of the National Flood Insurance Program in the 1970s and 1980s (Burby and French, 1981). The study found that local efforts to prevent floodplain development tend to be less successful when the supply of land outside of floodplains is scarce and/or local growth rates are high. Also communities with more fully developed floodplains, where efforts were likely to be less effective, tended to adopt the more aggressive programs.

Thus the timing of adoption of land use management tools for hazard mitigation makes a difference in effectiveness, as does the total effect of a locality's ongoing growth management system. If the local growth management system allows for sufficient land for development (land with adequate urban services and minimal environmental liabilities), then it is much more likely that restrictions on development of hazardous land will be effective. In other words, land use management for hazard mitigation is much less likely to be effective unless the overall development system approach in the locality is effective.

Finding a single summary measure for the comparative effectiveness of different tools has severe limitations. For example, Godschalk et al. (1989) undertook an evaluation of the effectiveness of local government mitigation practice in high-hazard coastal localities. They asked planning officials in 4 counties in Hawaii and 598 coastal localities with "velocity-zones" (or V-zones) for their perceptions of the effectiveness of their currently employed mitigation measures. Their findings generally confirm the findings mentioned earlier. The surveyed officials rated building and other structural and siting codes as more effective than land use review and control measures. Yet such summary measures of effectiveness have several limitations. Obviously, a single numeric rating given by a single respondent from each sampled locality tells us little about what effectiveness really means. A less obvious limitation is that such effectiveness ratings tell us more about what has been effective in the past, and very little about what may be effective in the future and under what conditions.

In reviewing for the National Earthquake Hazard Reduction Program the use of planning and land use management to reduce seismic hazards, Mader and Tyler (1993) adopted a qualitative approach to briefly summarize their judgments about the effectiveness of each of the major categories of tools. The authors of this book have extended Mader and Tyler's pragmatic approach to a larger array of tools as listed in Table 6-2. We also have extended the evaluation process by considering findings both from the completed research and from some still-ongoing research. The ongoing research projects, generally supported by the National Science Foundation, include those dealing with FEMA's section 409 mitigation planning process (Godschalk and associates), the possible role of taxation (Deyle and associates), applications of sustainable development principles (Berke and associates), integrating hazard management into local planning (Kartez and associates), use of seismic hazard maps and seismic safety plans by local governments (Olshansky, French), continuing research on the role of mandates and coercive versus cooperative approaches in improving the intergovernmental patchwork (Burby and May and associates), and research on enforcement of regulations (Burby and May and associates).

Two of the themes that occur frequently in the evaluation of different land use management tools in Table 6-2 are the need to extend existing local and state policies and tools to hazard mitigation objectives, and the need to link different tools together into a system in order to be effective. Both of these hazard mitigation implementation issues have only recently begun to be addressed in research.

TABLE 6-2 A Qualitative Assessment of the Effectiveness of Land Use Management Tools for Hazard Mitigation

Land Use Management Tools	Theoretical Effectiveness	Actual Effectiveness
Building Standards		
Special building standards for flood and seismic hazards (floodproofing requirements, or special seismic building codes)	Effective in elevating structures in floodplains to prevent building damages and widely used because of the National Flood Insurance Program. Seismic codes can effectively save lives and dramatically reduce (but not prevent) chances of building collapse. But seismic codes do not necessarily reduce overall damages.	Floodproofing does not reduce areawide flooding becauase, perversely, it encourages floodplain development. Seismic building codes are only very weakly enforced in central and eastern states and outside metropolitan areas at present.
Development Regulations		
Zoning ordinances	Can limit exposure of new development in hazard areas and protect natural values and functions not yet degraded by development.	Cannot mitigate losses to existing development and infrastructure. Can be subject to politically driven changes. Effectiveness of zoning is constrained to hazards for which risk varies geographically within a given jurisdiction.
Subdivision ordinances	The key point in land use management where damage can be reduced by design and by relocation of structures and lifelines to safer areas of subdivision.	Although more widely used than even zoning, subdivision regulation is not well tied to hazard mitigation aims in many areas.

Shoreline setback regulations	Can protect beaches and development from low-intensity coastal storms and flooding impacts, and prevent the need for sand replenishment.	Long-term limitations in that setbacks are set by mean high tide and not actual erosion conditions. That not only allows for catastrophic damages in extreme hurricanes, but also invites takings-style challenges when shallow lots preclude such setbacks. Too many variances given locally.
Special hazard area, fault-setback, or flood hazard area regulations	Regulations that combine features of zoning and building codes can be used to trigger more attention to mitigation options at the time of development.	Requires detailed information on the spatial extent and nature of hazard to be effective. Cannot determine the overall pattern of development, but can ensure that case-by-case mitigation opportunities are identified.
Dune protection regulation	Provides protection against storm surge flooding from low-intensity coastal storms. Few impediments. Generally accepted as wise practice in coastal areas.	Does not prevent development patterns vulnerable to catastrophic coastal storms.
Wetlands protection regulation	Prevents development of flood-moderating holding areas and also helps to clean urban storm run-off. Has been a fortunate coincidence that wetlands regulations may deal with floodplains, but hazards are tangential to wetlands laws.	Coastal and saltwater marsh regulations have been most effective when part of state and federal coastal management mandates of broader scope. Freshwater wetland regulation varies greatly from one jurisdiction to another. Filling of natural wetlands due to replacement provisions of no-net-loss laws may actually detract from flood reduction. Not well linked to hazard mitigation aims away from coastal storm zones.

continued

TABLE 6-2 *Continued*

Land Use Management Tools	Theoretical Effectiveness	Actual Effectiveness
Natural hazards identified in environmental impact assessments	Can be effective in changing the design and possibly overall intensity of of individual projects to mitigate hazards.	Like special hazard area regulations, cannot determine the overall pattern of development that produces hazard exposures. Only mitigates on a project-by-project basis. Can be reduced to meaningless routine in states with EIA/EIR mandates. Follow-up and enforcement of mitigation actions is often lacking.
Critical and Public Facilities Policies		
Capital improvements program (CIP)	Can be useful in steering development away from hazard areas by limiting availability of necessary urban services.	Not widely used for hazard mitigation. Jurisdictions often ignore their own CIPs. Not an effective tool by itself for any land use policy objectives.
Location of public facilities outside of hazard areas (to discourage development)	Can discourage or reduce the intensity of development in hazard areas.	Does not alter the basic spatial pattern of private development in hazard. areas. Rarely used for hazard mitigation purposes, and much more effective when linked with complementary land use regulations and tax policies.

Location of public facilities to reduce risk of damage to infrastructure	Protects lifelines in a disaster and reduces costs of public property damage, which are largely borne by federal government.	Does not alter the basic spatial pattern of private development in hazard areas. Local governments have low incentives to protect noncritical facilities because costs passed on to federal government.
Land and Property Acquisition		
Acquisition of open space/recreation/undeveloped lands for mitigation	Ownership of site is most effective means to manage it. Multiple objectives are possible (e.g., recreation, flood mitigation, and neighborhood redevelopment in Tulsa).	May be relatively meaningless as a mitigation device unless part of a management system for an entire hazard area. Most open space acquired for reasons unrelated to hazard mitigation and possibly in conflict with it (e.g., in Florida, lands not acquired if subject to erosion hazard). Again, coordination of objectives a problem. Expensive.
Acquisition of damaged buildings in hazard areas.	An important element of a community relocation effort.	Present use is very limited and effectiveness depends on what happens to acquired structures and subsequent rebuilding on- and off-site. Expensive, with very high demands for commitment and coordination.

continued

TABLE 6-2 *Continued*

Land Use Management Tools	Theoretical Effectiveness	Actual Effectiveness
Relocation of existing hazard area development to new site(s)	Removes risk to residents in the hazard area if no vulnerable rebuilding takes place.	Same limitations for building acquisition noted above. In addition, relocations require large investment in new site, with no assurances that former residents will move to relocated development. Timing is a problem because buyouts and relocation are not necessarily at the same time.
Acquisition of development rights or easements	Potentially very effective if funds are available and adequate authority (such as eminent domain) can be employed to target key sites.	Not used for hazard mitigation very frequently. Mitigation objectives are not linked to other acquisition programs for open space and growth management. Fee-simple acquisition tends to be emphasized more than partial rights/easements.
Transfer of development rights (TDR) away from hazard areas to safer locations.	Potentially very effective if there are suitable receiving areas for transferred rights and the program is mandatory, not voluntary.	Not used yet for hazard mitigation. Incentives not sufficient for using this complex tool for hazard mitigation aims alone.
Taxation and Fiscal Policies		
Preferential (reduced) taxation for open space or reduced land use intensity of lands in hazard areas	Important as a possible incentive for easements and other partial-fee transactions to limit development in hazard areas.	Has not been used for mitigation aims. Completely ineffective as a stand-alone tool. Requires state enabling legislation or extension of existing farmland and open-space laws to mitigation purposes.

Impact taxes or special assessments to fund the added public costs of hazard area development	Can shift costs of future public losses due to developing in hazardous locations back onto the developers and owners. Possible disincentive to vulnerable development.	Has not been used for mitigation, although many other public costs of of development are now collected from new development.
Information Dissemination		
Public information program	Better informed citizens and consumers can create a political constituency for hazard mitigation when they know about the location and magnitude of hazards.	Generally, these programs have a mixed record in building local political commitment for hazard mitigation. Targeted self-help programs that provide specialized information on floodproofing or residential earthquake mitigation have been more effective.
Construction practice seminars or builder/developer mitigation	Essential aspect of effective use of specialized codes and building standards. Can contribute to success of an overall multi-tool mitigation strategy.	Uneven education practices among states, code authorities and professional organizations detracts from the gains possible.
Hazard disclosure requirement in real estate transactions	Better informed real estate purchasers should create pressure for limiting some of the worst cases of new development in known hazard locations.	State-mandated disclosure as part of real estate sales has not been effective at all. Disclosure typically is perfunctory and is provided too late in the transaction to affect the purchase decision more than very marginally.

SOURCE: Based on a conference of the authors of this book, held at the Center for Urban Studies, University of North Carolina at Chapel Hill, March 22, 1996.

CONCLUSIONS

To carry out comprehensive plans, local governments use land use management tools. Logically, then, those governments that have the foresight to address hazard mitigation in their plans can adapt land use management tools to meet their hazard reduction goals. Indeed, many governments, at least since the early 1980s, have been using such tools for just that purpose. Over half of sampled local governments subject to floods, hurricanes, and earthquakes routinely use their zoning and subdivision ordinances to mitigate hazards to new development, and approximately half require use of hazard-resistant building standards. Depending on the nature of the hazard, even more tools are available: policies for location of public facilities, programs for acquisition or relocation of hazardous properties, fiscal policies, and information programs. Jurisdictions with a coherent strategy—and there are many—are able to build communities that will resist natural disasters, will recover quickly from them, and will last for many years with little cost to their inhabitants in dollars or lives. These are resilient, sustainable communities.

This chapter has reviewed the types of tools used, factors that can ease their adoption by local governments, and means of improving their local implementation. It has provided a realistic assessment of their effectiveness. Although local roadblocks are formidable, this chapter summarizes dozens of studies that show that many local governments have been successful in adopting land use management measures to mitigate hazards. Local governments that plan, create public awareness of the hazards they face, have policy advocates, and have adequate staff resources are most successful. In addition, it helps to have mandates and assistance from higher levels of government, previous experience with hazards, adequate community resources, a window of opportunity following a disaster, and a political culture that supports regulating private property for public ends. Governments must be careful in implementing programs, minding both the political and technical details. Some of the lessons from local experience show that communities must be both visionary and pragmatic. They need to be farsighted in gathering credible data, preparing maps, and managing land well before it is developed; but they must also be practical in using site-specific approaches, integrating hazard mitigation into their normal development review procedures, taking advantage of post-disaster windows of opportunity, and being prepared to purchase properties if necessary.

The entire list of potential land use management tools is not for every community. Each community and each local government must make choices, depending on the characteristics of the hazards they face and the political and economic qualities of the community. Use of the information in this chapter can help to inform these choices.

CHAPTER SEVEN

The Third Sector: Evolving Partnerships in Hazard Mitigation

ROBERT G. PATERSON

LAND USE PLANNING AND management are widely recognized by scholars as effective ways to prevent loss of life and lessen property damage from natural hazards, but many local governments have a poor track record in actually using them, as we noted in the previous chapter. Obstacles include deficiencies in commitment to implementing land use measures and shortfalls in local capacity. Several natural hazard researchers, however, suggest that nongovernmental groups (the so-called third sector) can play a significant role in creating pressure and support for land use measures (Futrell, 1986; Kusler and Larson, 1993; Lecomte, 1992). While third sector organizations are not newcomers to the field of natural hazards, their roles as agents or advocates for land use approaches to hazard mitigation have been limited.

This chapter explains how the third sector can play a wider role in promoting the use of land use measures to reduce losses from natural hazards. The following section describes the growth and significance of the third sector in policy making and explains why the third sector offers promise for promoting hazard mitigation. Issues surrounding the design of third sector partnerships are also explored. The chapter concludes by examining some of the more promising partnerships that policymakers and planners can pursue through the third sector.

THE PROMISE OF THE THIRD SECTOR

The third sector—also known as the nonprofit, nongovernmental, independent, or voluntary sector—encompasses everything from large-scale nonprofit institutions with paid professional staffs (e.g., the American Red Cross) to informal grassroots entities with no budget, no real legal status, and only good will and volunteerism as their principal resources (Van Til, 1988). From a global perspective, the United States is unique in the role that the third sector plays in articulating and mediating public policy and in providing public service (Carnegie Commission, 1993; Salamon, 1994). The explosive growth of the third sector over the last three decades springs from a variety of pressures, ranging from individual citizen demands, foundation and corporate support, and direct government action (Salamon, 1994). At present, there are more than one million nonprofit entities in the United States, and they continue to grow at a rate of about 30,000 a year (Clifton and Dahms, 1993).

Some scholars have argued that the recent rise of third sector influence is not a short-term phenomenon, but marks a revolutionary change in the importance of this sector relative to the government and market (Boulding, 1991; Gurin and Van Til, 1989; Ronfeldt and Thorup, 1995). The third sector has become more important in part because government's ability to act has become more constrained by ever increasing demands, legal rules, limited resources, and low public confidence, and because the for-profits are either not trusted or not willing to meet many pressing public needs. Lacking a profit motive and endowed with high levels of public trust, third sector entities have been thrust into a myriad of roles—catalyst for policy change, collaborator, and coordinator; adaptive and innovative deliverer of services; resource provider; and agent for local empowerment (Brown, 1991; Ronfeldt and Thorup, 1995; Uphoff, 1993). Third sector entities are often highly effective in these roles, and some researchers argue that third sector collaboration is an essential ingredient for solving seemingly intractable problems in government, business, and civil society (Ronfeldt and Thorup, 1995).

CONSIDERATIONS OF FUNCTION, SCALE, AND METHOD

Three basic design choices determine the efficacy of hird sector partnerships—(1) functions; (2) scale of operation (e.g., state, regional, or local); and (3) the method used to enlist the third sector in land use approaches to hazard mitigation. Third sector functions can include building *local commitment* to change by acting as a policy advocate and collaborative problem solver, building alliances for change, and easing coordination of the activities of citizens and government. The third sector also has been effective in building *local capacity* for change by acting as an innovative and flexible deliverer of services, a research and educational resource, and a financial supporter of local efforts. Table 7-1, on the following pages, elaborates on these functions and provides examples of how they can be employed effectively for hazard mitigation.

While third sector entities can perform a wide variety of functions, if the number of functions performed by any one entity becomes too large, conflicting responsibilities can diminish their effectiveness. The network of third sector entities that formed to promote the restoration of the Chesapeake Bay provides an excellent example of this situation (see Table 7-2). The Chesapeake Bay Foundation, the largest of the three nonprofits, serves an important role as a watchdog of public agencies and polluters. However, because of that role, it is not as effective in providing a neutral forum for education and policy dialogue and is limited in its ability to attract government and corporate financial support. The Alliance for Chesapeake Bay has assumed the neutral forum function, sponsoring workshops and conferences to raise awareness and support for restoration efforts at the household, community, and state levels. The Chesapeake Bay Trust's primary role is resource provider: it has charter provisions that restrict its range of operation to activities that have broad public support in order to avoid jeopardizing the willingness of private corporations, foundations, and private donors to contribute funds (U.S. Environmental Protection Agency, 1993).

The second critical choice is the scale of third sector operations: at what level should the functions be applied—national, state, regional, local or some combination thereof? Hazards researchers Berke and Beatley's (1995) recent adaptation of local institutional development theory (Uphoff, 1993) to the hazards context suggests that land use mitigation strategies are most likely to be successful when solutions are crafted, broadly supported, and implemented at the local level, with capacity-building support from higher levels of government and other or-

TABLE 7-1 Third Sector Functions That Build Local Capacity and Commitment for Land Use Approaches to Mitigation

Functions	Examples
Commitment Building	
Issue advocacy and mobilization of public opinion. The third sector is the locus of many of the most significant social movements in the U.S.: civil rights, the environment, and government reform (Van Til, 1988). It has a proven track record in its ability to lend legitimacy to an issue, to create political leadership, to increase the quality of debate, to generate political will, and to create an attentive public by raising awareness (Carnegie Commission, 1993; Hofner, 1990; Van Til, 1988; Wapner, 1995).	Insurance Institute for Property Loss Reduction, lobbying and support of mitigation strategies (Lecomte, 1992); Insurance Service Office, public awareness program to generate pressure on localities to improve building and zoning code effectiveness (Insurance Service Office, 1998); American Red Cross, mitigation initiative (Deutsch, 1996).
Policy form/collaborative problem solving. Third sector entities enjoy a higher level of public trust, on average, than government and private sector entities. They can create neutral, informal forums where power and status are not required for participation. These are settings where all stakeholders are encouraged to discuss their interests openly and to search collaboratively for solutions to satisfy all stakeholders' interests to the greatest degree possible. As independent nonprofits, third sector entities may be trusted to play a facilitating or mediating role in resolving conflicts or in attempting to build consensus on policy issues (Maser, 1995; Mawlawi, 1993).	In situations where scientific uncertainty is high and the risk of a hazard is contested, nongovernmental organizations (NGOs) can facilitate dialogue or mediate solutions (Carnegie Commission, 1993).
Coordination and partnership building. Because they have neither a commercial interest nor governmental status to protect, third sector entities have become the key brokers and intermediaries in establishing networks, alliances, and partnerships across government, industry, and communities (Brown, 1991; Gurin and Van Til, 1989).	Colorado Mitigation Council and Foundation coordinates state mitigation activities (Fred Sibley, State of Colorado, personal communication, 1995). Earthquake hazard reduction consortia coordinate on a multistate basis and support regional projects (Durham, 1993).

Capacity Building	
Trusted, innovative, and flexible service delivery. Third sector entities can try new ideas and services in ways that many public agencies may not because of statutory constraints or lack of knowledge. As independent nonprofits, they can try out new ideas too controversial for government to deal with and that the public would not entrust to profit-minded private entities. Given the lack of regulatory restrictions, the third sector can also adapt to changing circumstances and opportunities rapidly without concern for such issues as competitive bids or procurement regulations (Gurin and Van Til, 1989; U.S. EPA, 1993).	Standard-setting for public service professional associations, including the National Fire Protection Association, the Applied Technology Council, and Building Seismic Safety Council. Mitigation and recovery planning of hazard-prone areas (American Institute of Architects, 1991). Acquisition or donation of hazard-prone lands (Hocker, 1996).
Capacity Building	
Providing, attracting, and leveraging resources. Third sector entities entitled to IRS 501(3)(c) status are uniquely positioned to attract funding from a multitude of sources, including foundations, corporations, private donors, and government. They are increasingly obtaining funding through more traditional means such as subsidiary for-profit enterprises. Voluntary contributions of in-kind services also represent significant resources that can be leveraged for public purposes (Gurin and Van Til, 1989; Salamon, 1994; U.S. EPA, 1993).	Volunteer professional services for mitigation planning (American Institute of Architects, 1991). Specially created foundations to attract tax deductible contribu-tions from larger foundations and corporate giving programs.
Research, education, and information dissemination. Information is the primary currency of the third sector, and effective organizations become adept at creating and disseminating information to target audiences. Third sector entities play key roles in technology diffusion and information exchange at all levels of government and across all sectors of society (Boulding, 1991; Carnegie Commission, 1993; Gurin and Van Til, 1989; U.S. EPA, 1993).	Hazard mapping programs, conferences and symposia on mitigation; targeted publications and workshops to disseminate new technology and land use practice; model hazard mitigation plans and implementing ordinances.

TABLE 7-2 Third Sector Roles in Building Local Capacity and Commitment to Chesapeake Bay Restoration Efforts

	Chesapeake Bay NGOs		
Third Sector Functions	Alliance for Chesapeake Bay	Chesapeake Bay Trust	Chesapeake Bay Foundation
Capacity Building			
Information clearinghouse	X		
Policy, planning, and regulation			X
Implementation			X
Grant-making and funding support		X	
Commitment Building			
Forum for all interested parties	X		
Public and professional education	X	X	X
Research and monitoring	X		
Lobbying			X
Watchdog and litigation			X

SOURCE: Adapted from U.S. Environmental Protection Agency, 1993.

ganizations. A comparative case study of post-disaster mitigation planning by Berke et al. (1993) supports this theory. The researchers note that third sector entities were essential players in developing problem-solving capacity in local institutions and groups, linking community groups in common purpose, and finding support for mitigation from higher-level organizations (e.g., foundations, government, and corporations).

The nongovernmental organization (NGO) network dedicated to the Chesapeake Bay restoration is instructive on this design issue as well. For example, at the local level, all three organizations are effective in building local capacity and commitment to restoration efforts through public education programs. The Chesapeake Bay Trust also builds local capacity for restoration by providing grants to qualified community associations, civic groups, environmental groups, and local governments. The Alliance for Chesapeake Bay's networking activities and its volunteer monitoring program create a cadre of technically knowledgeable and committed citizens at the local level. Likewise, the Chesapeake Bay Foundation's 17 education centers have provided environmental field training to an estimated 33,000 students and have effectively advocated land use planning in numerous jurisdictions in the area of the Bay.

The Chesapeake Bay network also has been effective in finding and maintaining support from corporations, foundations, and the state and federal government. For example, the Chesapeake Bay Trust gathers support from key legislative, executive, and state agencies through its broad-based and powerfully placed 19-member board of trustees. The Alliance for Chesapeake Bay, a specialized networking entity, eases communications and gathers technical assistance and support from a broad range of entities, including the Environmental Protection Agency; foundations in Virginia, Maryland, and Pennsylvania; and large corporations. The Chesapeake Bay Foundation, with its more than 78,000 members nationwide, keeps Bay restoration efforts a high priority in government through sustained lobbying and litigation.

The third design choice—the method policymakers and planners use to enlist the third sector—includes the following options: (1) creating new third sector entities tailored to specific mitigation needs; (2) fostering an environment that spurs the independent formation of new entities by providing technical assistance, funding, and other forms of support; or (3) working with existing third sector entities whose interests are closely aligned with the goals of land use planning and management.

The direct approach, government creating new NGOs, is well established at all levels of government and in many policy arenas. There is considerable diversity in how these NGOs are created and in the nature of their relationship with government. Governmentally organized NGOs, or GONGOs, rely on significant levels of direct governmental financial support. They can be created through a variety of legal means including special acts, executive orders, or voluntary agreements. However, they must still seek out and qualify for Internal Revenue Service certification as 501(3)(c) organizations in order to enjoy many third sector benefits. In the Chesapeake Bay network example, the Chesapeake Bay Trust is a GONGO: it was established by a special act of the Maryland General Assembly and then sought and received a 501(3)(c) classification from the Internal Revenue Service. The American Red Cross is an example of another breed of governmentally sanctioned NGO known as a "QUANGO" (a quasi-nongovernmental organization). While the American Red Cross still enjoys the privileges and benefits of an independently governed nonprofit entity, its close ties to the federal government (it is congressionally chartered and plays a unique role in federal disaster relief) and substantial federal funding place it in a unique position, falling somewhere between the public sector and the third sector (Carnegie Commission, 1993).

Government policymakers and planners may also elect to operate in a more indirect fashion by creating an environment that has a high probability of fostering the formation of new third sector entities. Citizen participation scholars (Haeberle, 1989), community empowerment scholars (Prestby et al., 1990) and hazards researchers (Quarantelli et al., 1983) have identified a number of factors that are important for stimulating this kind of activism. For example, to mobilize interest and keep members active, grassroots organizations must be able to perceive and articulate a threat or problem that members have a real stake in solving (Haeberle, 1989). Government projects that disseminate historical accounts of community disasters, case studies of near misses that could have been disastrous, or even well-targeted community hazard mapping programs disseminated to the most at-risk local groups help create the prerequisite awareness needed for group mobilization.

Another important ingredient for grassroots activism is the belief that citizen action may be effective (Haeberle, 1989; Quarantelli et al., 1983). Thus, programs that acknowledge the concerns of grassroots groups and offer technical assistance or support in networking (e.g., a mitigation hotline, a clearinghouse information center, or a state hazard mitigation expert contact list) help spur activism. In addition, while most grassroots groups rely on volunteers and local expertise to accomplish their missions, financial support and assistance with organizational development (e.g., directions on how to incorporate and file for nonprofit status) help limit recurring problems such as participant burnout and disillusionment when early accomplishments are limited (Quarantelli et al., 1983; Di Smith, 1994). In the Chesapeake Bay network example, the Chesapeake Bay Trust is a designated funding agency that seeks to build both local capacity and commitment to restoration by allowing localities and NGOs to take on special projects and by helping local NGOs sustain their operations.

In many ways the last method of enlisting the third sector—collaborating with existing NGOs whose interests are closely aligned with land use approaches to hazard mitigation—may be the most important of all. Working with existing third sector entities offers a number of advantages over government formation of NGOs or fostering the spontaneous formation of third sector organizations. These advantages include avoiding start-up costs, capitalizing on existing contacts, enjoying immediate name recognition, and gaining instant public credibility. Moreover, given the growth and diversity of the third sector, this option offers a wide variety of opportunities for creating partnerships.

THIRD SECTOR PARTNERSHIPS: FUTURE CHOICES

Four of the more promising partnerships that government policy-makers and planners can pursue with existing NGOs to promote land use approaches to hazard mitigation include: (1) existing preparedness and response organizations within the hazards field; (2) environmental NGOs; (3) public service professional associations; and (4) universities and foundations. Each has promise, but none of them has been used widely. Here I note the possible role of each group and suggest ways in which government can collaborate more effectively with each of them. Next, governmentally created hazard mitigation NGO models are examined as well as alternative models of how government can enable grass-roots formation of hazard mitigation NGOs.

Preparedness and Response Organizations

Nonprofit organizations play an important role in disaster preparedness, response, and recovery in the United States. An umbrella organization known as the National Voluntary Organizations Active in Disasters (NVOAD) coordinates voluntary groups responding to disasters. Significant third sector relief organizations include the American Red Cross, the United Way, Church World Services, and the Salvation Army. Among these, the American Red Cross stands out as the largest non-profit in terms of institutional resources (spending nearly $1.5 billion on disasters between 1982 to 1992), scope of action (it has nearly 2,500 local chapters active nationwide), and importance to federal efforts (the Red Cross is the only nongovernmental member of the federal emergency response system).

While the Red Cross is best known for disaster preparedness and response activities, in early 1994 it announced a new "Disaster Mitigation Initiative" to work with government and private organizations. In May of 1994, the Red Cross Board of Governors adopted policy 2.6.5, which states: "The American Red Cross will advocate programs and legislation which mitigate disaster damage and loss of life, such as the *adoption of land use regulations, improved building codes and appropriate construction standards.* It will also advocate effective federal, state and local government programs that meet the recovery needs of disaster victims" (emphasis added; Deutsch, 1996). Its current mitigation plans include activities to build local commitment to land use strategies through awareness and education programs, as well as advocacy of local adoption and enforcement of building codes and land use rules in hazard-prone communities (Deutsch, 1996). A recent example of efforts along these lines is the *Next Big Earthquake* newspaper supplement that was disseminated to all households in the San Francisco Bay area. The newspaper supplement, sponsored by the American Red Cross, the United Way, and the U.S. Geological Survey (USGS), included a discussion on the role of land use planning techniques, the use of hazard maps, and additional references to regional planning approaches (USGS, 1994). Plans to build local capacity for hazard mitigation include helping disaster victims identify resources to fund mitigation projects (e.g., relocating and elevating homes from floods) and, as a last resort, direct funding of local mitigation efforts.

Because efforts of the Red Cross to foster mitigation are recent, it remains unclear how aggressively and effectively the new objectives will be pursued. Land use, which is significantly far afield of the American Red Cross experience, may present problems (e.g., lack of in-house expertise, staff, and/or chapter resistance; donor complaints; and donor-funding restrictions). Nevertheless, education and advocacy by the Red Cross can create a broad base of support for more effective hazard mitigation. Since many relief organizations enjoy strong public credibility and goodwill, these organizations seem especially well suited to act as ambassadors for land use approaches to mitigation. Federal financial support, technical support, and assistance in linking relief organizations with other NGOs having land use mitigation expertise may overcome some of the likely problems of staff commitment and capacity noted above.

Environmental Groups

Policymakers and planners may find further support for land use approaches to mitigation from organizations that advocate environmental sustainability. Hazard reduction and environmental protection are mutually reinforcing activities that often promote more sustainable communities (Berke and Beatley, 1995; Hamilton, 1992). The complementary relationships between hazard mitigation efforts, environmental protection, and, ultimately, sustainability become clear when consideration is given to how healthy natural systems often serve to protect communities from hazards and how land use strategies in turn often serve to keep those natural systems healthy. For example, natural coral reefs and barrier islands provide physical protection to low coastlines against damaging waves, tidal surges, and storm damage. Similarly, mangrove stands

The Loma Prieta Earthquake, 1989, San Francisco. The city's Marina district, built on 1906 rubble, sustained heavy damage in the 1989 Loma Prieta earthquake. The first story of this three-story building collapsed because of ground shaking and liquefaction. The second story also collapsed, leaving only the third story (NOAA).

and coastal forests help to dampen waves and high winds from tropical storms and lessen the impacts of tsunamis by absorbing part of the wave energy (Bender, 1995). Marshes and wetlands are also valued buffers for upland sites during hurricanes and coastal storms and provide natural storage for riverine flooding. Heavily vegetated or forested hillsides help to prevent mudslides and facilitate run-off infiltration which can reduce downstream flood stages.

While natural features can lessen the severity of hazards, many of these features face destruction because of their location in areas desirable for new development (e.g., at the water's edge or in places with a view). Land use planning and management can protect these natural features. Coastal construction setback lines serve to protect natural dune systems and barrier islands; land acquisition and transfer of development rights allow marshes, mangroves, and wetlands to stay in their natural state in perpetuity; and land management activities such as prescribed burns allow fire regime ecosystems to function more naturally. Development regulations can also limit the amount of development permitted on environmentally sensitive and hazard-prone land, minimize the level of vegetative disturbance, and prohibit uses of hazard-prone areas that could lead to unpredictable releases of hazardous substances when an earthquake, flood, hurricane, or another hazardous event occurs.

Developing relationships between groups interested in hazards and environmental management is not a new idea. In the early 1980s, the Conservation Foundation undertook a collaborative effort with federal and state governments to build local government capacity and commitment to conserve coastal resources and to prevent loss of life and property from coastal storms and hurricanes. In one instance, the Conservation Foundation, with support from state and local governments, helped Franklin County, Florida, develop a state-of-the-art shoreline protection strategy which includes the following elements: (1) guidance for site planning and structural integrity of development in high-hazard zones; (2) restoration of sand dunes degraded by random beach access and insensitive development; (3) guidance on the rate and amount of growth that barrier islands and flood hazard areas can safely accommodate, taking

SIDEBAR 7-1

Land Use Advocates and Partners: Land Trusts, Watershed Associations, and Professional Organizations

The Nature Conservancy (TNC) incorporated in 1951 for scientific and educational purposes. Its mission is to preserve the full array of species, communities, and ecosystems in the world by saving the land and water on which their survival depends. TNC has over $1 billion in assets, protects or owns over 6 million acres, and operates the largest nature preserve system in the world, with over 1,300 separate preserves. It accomplishes its mission through identification and mapping of rare species and communities, land acquisition and protection (gift, easement, lease, purchase, or other means), and land use planning and management (with the aid of 20,000 volunteers and 1,400 staff) (Murray, 1995).

TNC has a reputation as a collaborator that avoids confrontations while searching for levers such as interorganizational partnerships, volunteer action, and technological breakthroughs to advance its mission. One of its core prin-

into account environmental and evacuation constraints; and (4) guidance on land use policies to protect the ecological integrity of the Apalachicola Bay system (McCreary and Clark, 1983).

Examples such as the Franklin County, Florida, case are rare, but opportunities to create such relationships are perhaps greater now than before, as more and more environmental organizations adopt principles of environmental sustainability as part of their organizational mission. The challenge to policymakers and planners is to identify organizations with the greatest potential to advocate and support land use strategies.

Land trusts (a type of environmental organization) deserve special attention as promising partners in helping governments promote land use approaches to mitigation. A land trust is a nonprofit organization working toward the protection or conservation of important natural and/or cultural resources typically through purchase of land or conservation easements. Land trusts may be single purpose or pursue multiple objectives. These include preserving the environment, agricultural lands, forest lands, recreational lands, viewsheds, open space, scientific sites, ar-

---

ciples is that it is essential to involve other participants: no one institution possesses the full complement of talents, skill, knowledge, and interests to implement environmental protection programs. TNC worked with the U.S. Forest Service to create a cost-share program to bring private dollars to match government funds for specific conservation projects. It also has engaged over 5,000 volunteers in Northern California in riparian reforestation efforts, and has strong land use planning expertise as a result of its extensive mapping and preserve management programs. The conservancy will work with any legitimate group, whether government, NGO, university, or corporation so long as it shares similar goals (Murray, 1995). TNC does not take positions on public policy issues (with only few exceptions) so that it can continue to effectively interact with diverse organizations (Williamson, 1994), and it has a history of reaching out to different groups (Skolnick, 1993).

The Brandywine Conservancy was created in 1967 out of concern for the future of the natural and cultural resources of the rural Chadds Ford, Pennsylva-

continued

cheological sites, historic sites, and opportunities for affordable housing. Land trusts vary considerably in scale and sophistication. They range from major national environmental organizations such as the Nature Conservancy—whose combined assets, professional staff, and high-level contacts place its capabilities on par with many major multinational corporations (see Sidebar 7-1)—to small-scale local groups such as the Mendocino California Coastal Land Trust, which depend solely on part-time voluntary efforts to succeed (Land Trust Alliance, 1995; Murray, 1995).

Interest in land trusts is a relatively new phenomenon. Approximately half of all land trusts were formed within the last 15 years. The fastest growth is occurring in the Western and Mid-Atlantic regions (Land Trust Alliance, 1991). While exact figures are difficult to find, the best estimate is that the more than 1,100 land trusts in the United States have protected somewhere on the order of 6.2 million acres (Elfring, 1989; Hocker, 1996). Land trusts have significantly bolstered local government capacity to implement land use approaches to mitigation. For

SIDEBAR 7-1 *Continued*

nia, area outside of Philadelphia. After a short period of operation, the organization expanded its area of interest to include the entire Wilmington-Philadelphia region of Pennsylvania, with emphasis on protecting Brandywine Valley resources. Over the last 30 years, the Brandywine Conservancy has established itself as a respected advocate and supporter of environmentally responsible growth and resource conservation. The conservancy has protected over 25,000 acres through conservation easements and direct acquisition (Land Trust Alliance, 1995). Much of this land is in the floodplain and the conservancy also has promoted responsible floodplain management by helping 35 communities across 6 counties to develop land use controls that protect natural features (Kusler, 1982; Mantell et al., 1990). The conservancy also runs a land stewardship program that helps private and public landowners to develop short- and long-range plans for protecting and improving their farmland, waterways, and historic sites. The conservancy accomplishes its mission with a trained professional staff which includes planners, landscape architects, natural resource managers, and specialists in historic preservation and environmental engineering. As of 1995, the trust had 17

example, acquisitions by land trusts often do double duty by preserving sensitive riverine, coastal, mountain, and forested areas and by helping to prevent inappropriate development in hazard-prone locations. A recent analysis of the mission statements of the members of the Land Trust Alliance revealed that over 226 land trusts nationwide include coastlines and floodplains as priority protection areas (Paterson, 1996). However, acquisitions and easements for conservation are not the only avenues through which land trusts can serve hazard mitigation purposes. According to a 1990 survey by the Land Trust Alliance (1991), better than 67 percent of all land trusts also engage in education campaigns, 42 percent conduct or assist in community land use planning, and 40 percent actively lobby on land conservation issues within their service areas. Thus, land trusts may also play an important role in bolstering local commitment to land use strategies.

California's Greenbelt Alliance, Connecticut's Housatonic Valley Association, and Pennsylvania's Brandywine Conservancy have been particularly successful in spurring land use planning and management

---

full-time staff, 7 part-time staff, 3,300 dues-paying members, and an annual operating budget of $1 million (Land Trust Alliance, 1995)

The Charles River Watershed Association formed in 1965 to promote integrated watershed management in the Charles River Basin of Massachusetts. It participates in public studies and projects relating to the river and its resources, and promotes public awareness and education on river basin issues through conferences and publications (Platt, 1987). In addition to its efforts to improve water quality, enhance development of greenbelt and river-front parks, and increase public access to the river, the CRWA has been a strong supporter and participant in the development of a unique floodplain management program that combines a natural storage system of upland wetlands and marshes with downstream structural controls. The original plan called for purchase of full title or easements on 8,500 acres of wetlands that would store in excess of 50,000 acre-feet of water (comparable to a medium-sized reservoir). Approximately 57 percent of that total was actually protected (National Park Service, 1991). The re-

continued

that is sensitive to both environmental and natural hazard concerns (see Sidebar 7.1). All three organizations stand out as exemplars of what multifaceted regional-scale land trusts can accomplish through acquisition programs, grassroots organizing, education, technical assistance, fund-raising, interorganizational collaboration, and active lobbying (Greenbelt Alliance, 1996; Mantell et al., 1990). However, only about one in five land trusts nationwide enjoys the resource base and staff expertise of trusts like the Brandywine Conservancy (Land Trust Alliance, 1991). Efforts to enhance the capacity of existing land trusts to implement projects consistent with hazard reduction may be an effective avenue for promoting more widespread hazard mitigation activity.

Private watershed associations and conservation organizations perform similar functions as land trusts and offer comparable promise as partners for promoting land use approaches to hazard mitigation (see Sidebar 7-1). Although systematic information on these entities is lacking, efforts such as the Environmental Protection Agency's watershed protection approach (U.S. EPA, 1994) may substantially increase the

SIDEBAR 7-1 *Continued*

mainder of the floodplain is protected largely through local zoning controls. Nearly three-quarters of the localities within the Charles River watershed have floodplain regulations in place. The CRWA closely monitors proposed major developments in the watershed to safeguard the integrity of the floodplain management program (Platt, 1987).

The Upper Arkansas Valley Wildfire Council (UAVWC) and Foundation were created in 1992 and 1993 (respectively) to use land use planning, public education, and hazard reduction and response planning to reduce or prevent the loss of lives, property, and natural resources due to wildfire (UAVWC, 1993b). The forerunner of the UAVWC was a separate Urban-Wildland Interface Fire Committee in Lake County and Chaffe County, Colorado. The two counties combined their resources into a regional approach because of their proximity, and because they share similar lifestyles, attitudes, fuel types, weather patterns, topography, ignition sources, and values at risk. The impetus to these early urban-wildland committees came from local foresters and fire professionals whose awareness of and concern for wildland hazards in their own

number and potential of these entities. An exploratory investigation of seven watershed associations suggests that they have been particularly effective in promoting public awareness, stimulating the creation of grassroots constituencies, and providing neutral forums where complex policy issues can be debated and addressed creatively (Platt, 1987).

Clearly land trusts and watershed associations are just two examples within the larger pool of possible partnerships that can be explored with environmental organizations. Collaboration with major national environmental organizations that have extensive and highly active state and local networks, such as the Audubon Society and the Sierra Club, could have considerable impact on local mitigation efforts across the nation. Working with existing environmental networking organizations such as the Coastal Alliance would ease communications at all levels of environmental activism and across all sectors. Additionally, joining up with ongoing federal initiatives such as the National Park Service's Heritage Partnerships program (NPS, 1991) and the Department of Housing and Urban Development's empowerment community program offers prom-

communities were strengthened by participating in several wildland fire suppression efforts from 1988 to 1992 (several of which were out-of-state blazes) (Paul Summerfelt, Director, UAVWC, personal communication, 1996).

The UAVWC has been endorsed by both county governments but relies largely on volunteer efforts to undertake its mitigation efforts. The UAVWC Foundation [an IRS 501(c)(3)] was established to facilitate application for and receipt of grant money for mitigation efforts. Current mitigation strategies include: wildland fire hazard mapping; development of a "building permit kit" detailing prevention and mitigation measures for construction; new subdivision and building code changes; subdivision mitigation plans for existing high-hazard areas; and development of a dry hydrant plan for the Upper Arkansas Valley. The UAVWC has been nominated for several awards and has received requests for information from California, Texas, New Hampshire, Oregon, and Wyoming, which were interested in establishing similar councils (UAVWC, 1993a).

ise as well (all of these programs have embraced locally based planning as well as collaboration with NGOs as key ingredients of their success).

SCIENTIFIC, TECHNICAL, AND PROFESSIONAL ASSOCIATIONS

Scientific, technical, and professional associations have a long and distinguished record of creating and diffusing knowledge and thus helping to resolve or mitigate many social ills. Indeed, many researchers consider scientific, professional, and technical associations to be critically important for raising public awareness of natural hazards and for building support for mitigation strategies among policymakers (Alesch and Petak, 1986; Berke and Beatley, 1992a; Dynes, 1993b; May, 1991; Schulz, 1993).

There is some empirical evidence supporting the importance of these organizations in local policy development. For example, in the early 1980s, Bingham and his colleagues (1981) tested a linear process model of the influence of professional associations on innovation by local governments. They evaluated the impact of 15 public service professional associations on the speed with which innovations were adopted by a random sample of 323 cities. Virtually all professional associations received favorable evaluations by city officials in terms of their effectiveness in increasing awareness and providing data that fostered the adoption of innovations. However, all but 2 of the 15 professional associations received low ratings on their usefulness in helping line professionals and executives foster a climate of acceptance for innovations (which led Bingham and colleagues to suggest that this is an area where professional associations should increase their capabilities).

There is also some empirical evidence pointing to public service professionals and professional associations as important sources of information about hazard mitigation (see Sidebar 7-1). For example, Bolton and Orians' (1992) study of mitigation efforts after the Loma Prieta earthquake in the San Francisco Bay area identified professional associations as among the top two sources that provided the most useful technical assistance to local governments both before and after the earthquake. Alesch and Petak's study (1986) of seismic mitigation efforts in three California communities led them to propose (among other things) that the probability that mitigation measures would be adopted is in direct proportion to the extent that there are credible, persistent inside policy advocates (e.g., line professionals and executives) and that professional associations are a primary vehicle for communicating innovations in hazard mitigation among jurisdictions. Gori's analysis (1991) of the U.S. Geological Survey capacity-building experiment, involving the placement of geologists at the local level, shows that scientific professionals can be effective in influencing local agendas, increasing awareness of hazards, and easing implementation of mitigation measures. Furthermore, Schulz (1993) in summarizing studies on the transfer of earthquake loss reduction information noted that the most successful use of a product results from a personal contact between the information provider and the target user, and that the most effective programs are institutionalized in some manner, or they establish linkages with professional societies and organizations.

While there are a great many partnerships that could be pursued to foster land use approaches to hazard mitigation, public service professional associations which typically serve local government are an obvious target (see Table 7-3). They offer a number of advantages, including: high participation rates among local line and executive professionals and cities (which translate into multiple channels through which to diffuse mitigation messages), high credibility among local government decision makers, and the potential of expanding their organizational focus to promote land use strategies under the guise of their public service function (Bingham et al., 1981; Campbell, 1988; Waldo, 1973).

There are several examples at the federal level of efforts to engage these organizations. For example, the American Institute of Architects and the American Planning Association have been able to increase hazard mitigation awareness and professional competency through FEMA-funded seismic design workshops, educational development grants, and hazard mitigation research. In addition, FEMA also has worked with the National Sheriffs' Association to incorporate disaster preparedness materials into the National Neighborhood Watch program and worked with the National Emergency Management Association to create emergency planning guidelines for utilities and businesses. However, these efforts tend to be episodic at best and have had limited impact in terms of the total number of professionals contacted; have failed to be multi-hazard in focus; have not fully addressed land use mitigation alternatives; and have reached just a few of the professional groups that could influence local land use decision making.

The key question confronting policymakers and planners is how to better work with these highly credible and seemingly effective associations. Given that coordination and cooperation among professional as-

TABLE 7-3 Public Service Professional Associations Impacting Local Government Policy Making

Group	Title
Chief Executive Associations	American Society for Public Administration
	International City/County Management Association
	National Association of County Governments
	National League of Cities
	National Municipal League
	U.S. Conference of Mayors
Line Professional Associations	American Institute of Architects
	American Institute of Certified Planners
	American Planning Association
	American Public Health Association
	American Public Works Association
	American Society of Civil Engineers
	American Society of Landscape Architects
	American Society of Professional Engineers
	Government Finance Officers Association
	International Association of Chiefs of Police
	International Association of Fire Chiefs
	National Emergency Management Association
	National Institute of Municipal Legal Officials
	National Recreation and Park Association
	National Sheriffs' Association

sociations working on hazard reduction has been less than ideal to date (May and Stark, 1992; Schulz, 1993), perhaps a more sustained and coordinated approach would produce better results. For example, a formal hazard mitigation coordinating or umbrella organization could be used to create forums where representatives of professional associations could learn about federal, state, and regional mitigation priorities, while at the same time allowing government officials to better understand the organizational capabilities and synergies that may be possible by working in partnership with a variety of professional associations.

One of the more promising areas to spur greater collaboration among public service professions is within the design professions (e.g., architects, planners, urban designers, and landscape architects). Design professionals can have a profound influence on a community's exposure to hazards through their own professional practice and the lobbying activities and educational programs of their professional associations. Collective efforts among these professions to delineate "best mitigation practices" (encompassing a land use planning approach) or to establish principles for "disaster-resistant community design," could lend considerable legitimacy to land use efforts to reduce susceptibility to hazards. In addition, a number of new hazard-related policy networks could result from these forums, as well as new information dissemination resources to speed the transfer of knowledge about effective land use techniques. While perhaps not easy to achieve, the likely benefits of a broad-based multidisciplinary coalition of public service professionals certainly warrant further experimentation.

Universities and Foundations

Universities have made important contributions to efforts to reduce losses from natural hazards, but they could expand their influence further in the area of hazard mitigation. Universities contribute information to policy debate, influence public opinion through a variety of educational and information channels, and shape the intellectual focus and technical training of professionals and others who operate government and affect the built environment (Carnegie Commission, 1993). A number of universities support specialized hazard research centers which serve as clearinghouses on mitigation techniques, provide symposia and conferences, and offer hazard-related education and training. However, Schulz (1993), in a review of hazard reduction research efforts, argues that these centers, and that universities more generally, should expand their efforts to include training and education of local officials and other groups that ordinarily do not seek out hazard mitigation information but who are central to policy change.

Universities could also do more, as noted by Schulz (1993) and other investigators (e.g., Ender and Kim et al., 1988; Havlick and Dorsey, 1993) to increase the capacity and commitment of design professionals to use land use strategies to reduce losses from hazards. At present, hazard reduction curricula are not well integrated into undergraduate and graduate design profession programs. For example, a recent survey of graduate planning programs found no courses specifically focusing on natural hazards and hazard mitigation (Havlik and Dorsey, 1993). An earlier investigation by Miller and Westerlund (1990) identified only three courses across 57 graduate planning programs in the United States and Canada that specialized in hazard mitigation. While there have been some recent efforts to address this shortcoming, effort is required in schools, universities, and professional certification programs to ensure a future constituency for mitigation and a competent population of professional staff to implement mitigation measures (Schulz, 1993).

Foundations also could play a stronger role in promoting hazard mitigation. Foundations motivate, organize, and finance a variety of institutions that bring information and affected stakeholders together; engage universities and think tanks in policy-relevant research; fund natural experiments to evaluate new but untested ideas; and create forums where experts and citizens can collaborate to resolve problems (John A. Smith, 1989). However, foundations have yet to endorse mitigation of natural hazards as a priority. According to the Foundation Center's National Guide to Funding series, which combines information from five major grant-giving databases, less than one-half of 1 percent of all community development giving was targeted for disaster-related issues in 1993–1994 (Foundation Center, 1995). This amounted to just under $2 million, which is a relatively small amount when compared to the total of $635 million in grants awarded by the approximately 2,500 largest private, corporate, and community foundations in the United States. Furthermore, of the 44 disaster-related grants awarded in 1993–1994, virtually all of the giving was relief aid for recent disasters (see Table 7-4). Only two projects focused directly on preventing future losses through mitigation, and just one had a strong land use emphasis. That project, supported by the Knight Foundation, involved a series of post-hurricane design charettes and conferences to create plans for the redevelopment of 28 communities over a 140-square-mile area. The charettes and the planning process considered several ways to minimize flood and wind damage from the next hurricane (Vonier, 1993). Policymakers, planners, and perhaps newly created hazard mitigation nonprofits (see following section) can alert foundations to the benefits of land use strategies to reduce losses from natural disasters.

CREATING AND ENABLING THE FORMATION OF NEW THIRD SECTOR ENTITIES

Federal, state, and local policymakers are not newcomers to creating or sponsoring new third sector support organizations. Indeed, in the hazards context, most of the existing earthquake consortia, including the Central U.S. Earthquake Consortium, the Western States Seismic Policy Council, and the New England States Earthquake Consortium, rely largely on FEMA sponsorship to cover significant portions of their expenses (Durham, 1993). These organizations foster and coordinate

TABLE 7-4 Examples of Disaster-Related Giving, 1993–1994

Disaster	Foundation	Grant Purpose
Hurricane Iniki	Hawaii Community	Repair damage to foundation facilities
	American Express Foundation	Repair damage to facilities
Hurricane Andrew	Knight Foundation	Community charette for recovery
	Ryder System Charitable Foundation	Repair damage to low-income homes
Northridge Earthquake	Weingart Foundation	Repair damage to facilities
	California Community Foundation	Repair damage to facilities and aid those made homeless
	Ahmanson Foundation	Repair damage to low-income homes
Mississippi Floods	Needmor Fund	Relief aid
	McKnight Foundation	Community organizing for relief and recovery, needs assessment for recovery

SOURCE: Foundation Center, 1995.

implementation of hazard- and risk-reduction policies through symposia, conferences, workshops, demonstration projects, and publications (Durham, 1995). However, their focus is narrow in that they only address earthquake hazards, and their multistate focus limits the extent to which they can affect local-level mitigation efforts.

A better model is California's Bay Area Regional Earthquake Preparedness Project (BAREPP). BAREPP was created through a cooperative agreement between FEMA and the State of California in 1984. The agreement set the following objectives for the project: "(1) promoting comprehensive earthquake preparedness actions by local jurisdictions, volunteer agencies and associations and the private sector; (2) providing planning assistance and coordination in the development of improved regional response capabilities for predicted and unpredicted major earthquakes, including programs of test and exercises; (3) providing technical and planning assistance to local jurisdictions in the development and implementation of programs of hazard mitigation and prevention to reduce earthquake vulnerability; and (4) establishing a local incentive grant program to promote demonstration projects in comprehensive earthquake preparedness" (Orians and Bolton, 1992). A policy advisory board, consisting of major organizations and groups with an active interest in improved earthquake safety, consulted and advised the project staff on goals, objectives, and policies.

BAREPP worked hard to gain visibility and credibility within the communities it serves (Orians and Bolton, 1992). A survey of Bay area jurisdictions found that BAREPP was one of the two most often used and valued sources of information and technical assistance for both pre- and post-earthquake mitigation planning (Bolton and Orians, 1992). Furthermore, Eisner (1991) argued that BAREPP's work in the Bay region had a significant positive influence in reducing losses in the 1989 Loma Prieta earthquake because local governments knew what to do, many had mitigation programs in place before the earthquake, trained staff responded as planned, and hazardous structures were identified well in advance.

SIDEBAR 7-2

A State-Level GONGO and a National NGO Outreach Model

The Colorado Natural Hazard Mitigation Council and Foundation. Since its inception, the council has supported over 100 projects ranging from hazard-vulnerability mapping to post-disaster mitigation projects. The council is composed of almost 300 appointed, ex officio, and volunteer committee members. Thirty-five members and a chairperson are appointed directly by the governor. The council has six hazard subcommittees that give in-depth attention to mitigation of the following conditions statewide: severe weather, drought, geologic hazards, dam safety, wildfire, and floods. By far the greatest portion of the council's work relies on volunteer efforts and collaboration with other voluntary and governmental agencies. For example, in 1994, the council collaborated with the National Civilian Community Corps to provide a post-disaster mitigation workshop and post-disaster mitigation field work in the community of Lyons, Colorado (Fred Sibley, Office of Local Government Affairs, State of Colorado, personal communication, 1995). The establishment of the Colorado Natural Hazard Mitigation Foundation in 1991 [a 501(3)(c) certified organization], gave

It is important to note that while BAREPP was a promising regional model suitable for replication or adaptation in other states (perhaps on a broader multi-hazard basis), it was *not* a third sector entity. While it would have been quite easy for BAREPP to be formed as a third sector entity under the original agreement (and it could have become one), it operated as a project controlled by the California Governor's Office of Emergency Services and essentially went out of business when state resources were withdrawn.

Another model that encompasses broader hazard mitigation objectives is a state-level innovation—the Colorado Natural Hazard Mitigation Council (CNHMC) and Foundation (see Sidebar 7-2). The Colorado Natural Hazard Mitigation Council was created by executive order of the Colorado governor in 1989 in order to evaluate, prioritize, and implement hazard mitigation projects. This council, the first organization of its kind in the country, coordinates natural hazards research and loss reduction strategies throughout the state of Colorado and integrates the plans and resources of local, state, and federal agencies (Colorado

the council an additional vehicle for forming coalitions, attracting foundation support, and other public-private partnerships to fund recommended hazard mitigation projects (Colorado Office of Emergency Management, 1993).

The National Trust's Main Street Program. The National Trust was created by congressional charter in 1949 to encourage public participation in the preservation of historically and culturally significant sites, buildings, and objects. Since its inception, the National Trust has expanded its mission to serve as a central source of information on preservation technology, a forum for the exchange of ideas, a funding source for preservation planning, and a direct provider of services for preservation planning. In 1977, the National Trust launched three Main Street demonstration projects in the Midwest to find ways to preserve and revitalize small downtowns. Over a three-year period, it developed a four-point strategy that has been successfully replicated in over 1,100 towns and cities in 40 states since the program expanded to a full-scale operation in 1980.

The approach is based on community self-determination and gradual transformation, with small changes in the early years as revitalization efforts strengthen local capacity to tackle more complex challenges. The four points of the strategy,

continued

Office of Emergency Management, 1994). It provides information and technical assistance to local governments and individuals, identifying specific mitigation measures for a given area, and then assists in implementation. Volunteer committee members, who are challenged to obtain funding and in-kind support for their projects, accomplish the bulk of the council's work. In 1991, the council created the Colorado Natural Hazard Mitigation Foundation to provide a funding conduit for its efforts and to make the council more competitive for federal, corporate, and foundation support. The council has accomplished a great deal in its short existence (over 100 projects), and in 1996 it instituted a re-engineering process to ensure its continued effectiveness (Fred Sibley, Office of Local Government Affairs, State of Colorado, personal communication, 1995). The council and foundation present a promising model for other states to consider because of the multi-hazard mitigation emphasis and the excellent level of voluntary participation experienced to date.

A highly successful national-level model that has inspired widespread local activism and enhanced local planning capabilities (but that falls outside the hazards context) is the Main Street program of the National

SIDEBAR 7-2 *Continued*

on which work is performed simultaneously, are: (1) economic restructuring—strengthening the downtown's existing economic base while gradually expanding it; (2) design—improving the downtown's physical environment, rehabilitating historic buildings, building architecturally sensitive buildings, and ensuring that streets, sidewalks, signs and lighting, and other elements function well and support the overall design of the commercial district; (3) promotion—marketing the downtown's assets to residents, visitors, and investors; and (4) organization—building collaborative partnerships among a broad range of public and private organizations, agencies, businesses, and individuals.

To introduce the program to other communities, in 1980 the trust created a National Main Street Center, which includes a revitalization learning center and information clearinghouse. It has served as a model for similar programs in Canada, Venezuela, the United Kingdom, Australia, and New Zealand. In 1995, the program expanded to include neighborhood commercial districts as well as small and medium-sized city downtowns. By the end of 1993, for every dollar a community spent on its Main Street program, it leveraged an additional $25 in new investment (on average) (Keith Smith, 1995).

Trust for Historic Preservation. The Main Street program provides an excellent example of how a large, governmentally organized nonprofit can successfully expand its mission and operations to become an effective policy advocate and capacity builder at the local level (see Sidebar 7-2). The National Trust's Main Street coordinators provide technical support and other services to local voluntary preservation efforts. What makes this model especially interesting from a land use perspective is how Main Street coordinators place as much emphasis on planning what can be changed as on planning what must be preserved. Preservation is simply one component of a local Main Street revitalization plan, albeit an important one. Perhaps this model can work equally well in hazard-prone communities. However, as in the Main Street model, hazard mitigation would need to be linked to other important community development objectives.

Another promising model is New York State's Hudson River Valley Greenways Community Council and Conservancy. The New York state legislature created the council so that it can operate as a designated regional planning agency (it has official status as a governmental planning entity), as a public benefit corporation (to fulfill nonprofit implementation efforts), and as a 501(3)(c) (so it can attract funding from foundations and corporate giving programs). The Council and Conservancy created a model communities program that aims to foster a voluntary regional planning compact across 10 counties to protect the assets of the river valley. Ten model community planning projects were under way by 1996, involving 23 communities. The greenway's planning process builds local planning capacity by supporting the creation of planning committees, developing community planning profiles, and subsequently forming a vision based on several public meetings (Diamond and Noonan, 1996). Much of the planning focuses on the Hudson River, and floodplain management considerations are interwoven into the planning process. Strong state statutes that allow prohibition of most uses in the 100-year floodplain support planning for the floodplain. Interest in floodplain management issues took on greater urgency in 1996 when the Hudson River Valley suffered its worst flooding in over 70 years.

Both the Main Street and Hudson Valley model communities programs aim to stimulate grassroots activism and enhance local problem-solving capabilities by building local capacity and commitment to land use planning processes. In both cases, the program's primary planning emphasis—historic preservation and environmental conservation, respectively—is embedded within the larger framework of community planning. This allows stakeholders within communities to balance a number of competing community objectives, and to craft locally acceptable, creative solutions to meet as many of those objectives as possible.

CONCLUSIONS

Implementation of land use strategies for hazard mitigation has been difficult. Lack of public understanding, professional support, and political leadership have been significant barriers. The third sector can help overcome these obstacles through its ability to mobilize public and political support, shape public opinion, attract diverse funding, and leverage scarce resources. The third sector has been especially effective in stimulating, supporting, and sustaining community-based planning programs, but these efforts have given little attention to hazard mitigation. Several promising third sector partnerships to expand support for land use approaches to hazard mitigation have been explored in this chapter. They include: working with existing NGOs that have already started to branch out into the hazard mitigation field; attempting to build support for mitigation in cooperation with NGOs that have complementary missions (such as environmental organizations and public service professional associations); and creating or enabling new hazard mitigation entities. Policymakers and planners may replicate or adapt these models to suit their specific mitigation goals. However, because of limited third sector participation in the hazard mitigation field and lack of evaluative information on the third sector in general, it is difficult to say which options will prove most effective. For those reasons, past experiences and current experiments with differing NGO–government structures should be evaluated further to identify ways to leverage the best and minimize the worst aspects of these relationships. While the challenges to implementing greater use of land use strategies are formidable, the potential rewards certainly warrant greater experimentation with partnerships between government and the third sector.

PART THREE

Looking to the Future

CHAPTER EIGHT

The Vision of Sustainable Communities

TIMOTHY BEATLEY

S WE CONVERGE ON THE twenty-first century, sustainability represents the only viable paradigm in which to place land use policy and planning, and indeed environmental and social policy more generally. What sustainability means is somewhat ambiguous and controversial, but the vision of sustainability is a powerful and compelling one. It is at once a moral statement about how we should be living on the planet, and a description of the social and physical characteristics of the world as it should be. Any discussion of land use policy and planning, then, must be placed within a broader framework of sustainability. Land use planning to reduce natural hazards is ultimately and fundamentally about promoting a more sustainable human settlement pattern and about living more lightly and sensibly on the earth. As noted in Chapter 1, the increasing severity and impact of natural disasters suggest a serious breakdown in sustainability, and that indeed we are not living in a very sustainable way on the planet.

This chapter presents a vision of sustainability in the context of the reduction of natural hazards. It begins by discussing the history of the idea and the different meanings and definitions that have emerged over the years. Next, the chapter discusses the connection between sustainability and natural hazards and identifies recent examples of how planning to foster hazard mitigation has been tied to sustainability. This chapter then extracts and presents a series of general principles of sustainability. These principles describe what the vision of sustainability implies in more detail; it can serve as a guide to more specific local and regional initiatives. The final section discusses specific steps in planning for sustainability, as well as tools, techniques, and methods for implementing the vision of sustainability.

THE MANY MEANINGS OF SUSTAINABILITY

To many, sustainability and sustainable development are just the latest buzzwords to make their way into the planning field—another set of fashionable phrases. There is no question that planners use these terms increasingly to describe what they do and what their professional mission is. A common criticism, however, is that meanings of the terms are not immediately obvious: *sustainability* and *sustainable development* require definition and elaboration, as do terms like *freedom*, *justice*, or *quality of life*.

We have a sense that sustainability is a good thing (and that being unsustainable is a bad thing), but will we know it when we see it? It is likely that this ambiguity will remain with us, but, in my view, the very fact that we are questioning what is or is not sustainable and exploring what the idea means and calls for is a very positive sign. It opens opportunities for critical dialogue and serves as an important catalyst for thinking clearly and systematically about the future planners can help to bring about.

Sustainability is the root term, from which stems its use as a modifier (e.g., sustainable development, sustainable forestry). A typical dictionary definition of *sustain* might include the following: to keep in existence, to maintain or prolong, to continue or last. Sustainability finds many of its roots in biology and ecology and specifically in the concept of ecological carrying capacity—the notion that a given ecosystem or environment can sustain a certain animal population and that beyond that level overpopulation and species collapse will occur (see Kidd, 1992). Central, then, is the idea that certain physical and ecological limits exist in nature that, if exceeded, will have ripple effects that bring population back in line with capacity.

The meaning of sustainability is perhaps clearest when applied to renewable resources such as ocean fisheries, forests, groundwater, and soils. Terms such as *optimal sustainable yield* have been incorporated explicitly into, for example, U.S. fisheries management law. The growing advocacy for sustainable use of a variety of renewable resources—sustainable forestry, sustainable fisheries, sustainable agriculture—is premised on the idea that these resources can be used and harvested in a manner that allows them to renew themselves and that preserves their long-term productivity. Sustainability is also a useful concept in planning the use of nonrenewable resources, the waste-assimilative capacities of the earth, and the natural services provided by the environment (e.g., climate regulation; see Jacobs, 1991).

The use of the term *sustainability* in environmental planning and policy circles is relatively new. It began appearing in the literature in the early 1970s and emerged as a significant theme in the 1980s, when sustainability was embraced by such nongovernmental organizations as the Worldwatch Institute and World Resources Institute, governmental organizations such as the U.S. Agency for International Development, and a number of international study groups. The term of preference became *sustainable development*, which focused on how human interventions—especially international development programs and projects—failed to respect the integrity of the natural systems in which they were sited. Projects of international development agencies such as the World Bank were severely criticized as the antithesis of sustainable development.

There have been a number of attempts to define sustainable development formally, especially in the last decade. Perhaps the most frequently cited definition is that put forth by the World Commission on Environment and Development—also commonly known as the Brundtland Commission (World Commission on Environment and Development, 1987). The WCED defined sustainable development as that which "meets the needs of the present without compromising the ability of future generations to meet their own needs" (WCED, 1987, p. 8). More recently the National Commission on the Environment has defined sustainable development as "a strategy for improving the quality of life while preserving the environmental potential for the future, of living off interest rather than consuming natural capital. Sustainable development mandates that the present generation must not narrow the choices of future generations but must strive to expand them by passing on an environment and an accumulation of resources that will allow its children to live at least as well as, and preferably better than, people today. Sustainable development is premised on living within the Earth's means" (National Commission on the Environment, 1993, p. 2). These definitions share an emphasis on certain important concepts and themes at a general level. They stress the importance of living within the ecological carrying capacities of the planet, living off the ecological interest (protecting the ecological capital), and protecting the interests of future generations. They envision a society that "can persist over generations, one that is farseeing enough, flexible enough, and wise enough not to undermine either its physical or its social systems of support" (Meadows et al., 1992, p. 209).

Application of concepts of sustainability to cities, towns, and settlement patterns has been an even more recent phenomenon, yet there has been an explosion of thinking, writing, and practice attempting to make this connection in recent years (e.g. see Aberley, 1994; Beatley and Brower, 1993; Blowers, 1993; Newman, 1996; Rees and Roseland, 1991; Roseland, 1992; Van der Ryn and Calthorpe, 1991; and many others). There are a variety of labels used including sustainable cities, sustainable urban development, and sustainable communities, among others (I will tend to use the latter in this chapter). All take as their beginning assumption that the current patterns and forms of urban development are not sustainable in the long run.

Common attributes of sustainable communities frequently identified in the literature, in professional discussions, and in practice include the following: compact, higher-density development, and the more efficient use of land and space; the "greening" of communities with greater emphasis on trees, parks, and open space; an emphasis on redevelopment of underutilized urban areas and on infill development; greater emphasis on public transit, and creating mixed-use environments which are more amenable to walking and less dependent on autos; and energy and resource conservation and low pollution, among other qualities. Increasingly, the resilience of a community to natural disasters is being added to this list of central qualities, as discussed in detail below.

Examples of sustainability initiatives at local and regional levels are increasingly common, with cities like Seattle and Chattanooga organizing their planning and redevelopment efforts around sustainability. The Habitat II meeting in Istanbul in 1996 had as a major theme sustainable cities, and the recent report of the President's Council on Sustainable Development (1996, Chapter 4), *Sustainable America*, gives considerable attention to the importance of sustainability at a community level.

In addition, however, a number of states also have undertaken studies, appointed commissions, or convened conferences aimed at taking stock of the sustainability of land use and development patterns (including, for example, Florida, North Carolina, and Pennsylvania). In Minnesota, the result was a report (Minnesota Environmental Quality Board, 1995) which identifies weaknesses and ways in which development is not occurring in a sustainable way, and a set of principles, goals, and guidelines for how local sustainability can be advanced.

There are strong reasons to believe that current growth and development patterns are not sustainable in the long term. Contemporary patterns of land use do not appear to acknowledge the fundamental fact that land, air, water, and biological diversity are finite. Land use patterns are extremely wasteful, economically expensive, and ecologically damaging. There is also reason to question the social sustainability, livability, and quality of life of the communities being created. (For telling critiques of American patterns of land development, see Calthorpe, 1993; Kunster, 1993; and Langdon, 1994.)

SUSTAINABILITY AND NATURAL HAZARDS

Exposure and vulnerability to natural hazards is an important dimension of the sustainability of cities and communities. Natural disasters dramatically illustrate the ways in which contemporary development is not sustainable in the long run. The unprecedented levels of property damages experienced in recent years, discussed in Chapters 1 and 2, are the clearest indication of the unsustainability of human settlements.

There has been considerable acknowledgment in recent years that sustainability and sustainable development imply efforts to create and maintain communities that avoid or mitigate natural hazards. Within the sustainability literature increasing attention is paid to natural hazards. *Agenda 21*, the action agenda adopted at the 1992 Rio Summit, gives considerable attention to reducing natural hazards, and clearly includes hazard reduction and avoidance in the definition of sustainable human settlement patterns (United Nations, 1992). Chapter 7 of the report proposes a number of pre-disaster planning and post-disaster reconstruction planning activities including "redirecting inappropriate new development and human settlements to areas not prone to hazards" and supporting efforts at "contingency planning, with participation of affected communities, for post-disaster reconstruction and rehabilitation" (United Nations, 1992, p. 61–62).

The recent report of the President's Council on Sustainable Development gives considerable attention to natural hazards, and includes as a specific action agenda item the identification and elimination of "government subsidies, such as subsidized floodplain insurance and subsidized utilities, that encourage development in areas vulnerable to natural hazards" (1996, p. 99). Increasingly, the literature and the debate about sustainability encompasses natural hazards (e.g., see also Berke, 1995; Geis and Kutzmark, 1995).

Similar connections are being made by the natural hazards community. One of the strongest expressions of the connection between sustainability and natural hazards can be seen in the recent report of the National Science and Technology Council (1996). The need for society to move toward greater sustainability is a major and recurring theme in its report, *Natural Disaster Reduction: A Plan for the Future*. The report suggests, for instance, that an additional criterion be added to what sustainable development means and requires (in addition to economic growth, environmental protection, and sustainable use of ecological systems): "There is . . . a fourth criterion of equal importance: sustainable development must be resilient with respect to the natural variability of the earth and the solar system" (National Science and Technology Council, 1996, p. 2). This natural variability includes such forces as floods and hurricanes and shows that much economic development is "unacceptably brittle and fragile" (p. 2).

Current patterns of land use have much to do with why development is "brittle." Many of the places experiencing the highest growth rates, and the most sprawling development patterns, are also subject to severe natural hazards. Consider the level of development in southern California, development along the Wasatch front in Utah, development in the Outer Banks of North Carolina or the Florida Keys, or redevelopment of the Oakland hills, as Platt points out in Chapter 2. Community land use patterns are clearly not sustainable if they allow or encourage the exposure of people and property to significant risks from natural hazards, and if alternative settlement patterns are available that would avoid such exposure.

Contemporary land use patterns and practices contribute to placing people and property at risk in both direct and indirect ways. Obviously, human settlement patterns are not sustainable when people and property are placed directly in harm's way (e.g., growth in close proximity to ocean fronts subject to storm surge and shoreline erosion, construction in high-risk landslide areas or wildfire zones, etc.). But breakdowns in sustainability also occur in more indirect ways, for instance when land use patterns and practices serve to undermine the ability of the natural environment to absorb hazards or generally alter important natural systems. The creation of dramatically greater areas of impervious surface (road surfaces, parking lots, buildings), for instance, can lead to increased downstream flooding. Constructing seawalls, floodwalls, or levees may serve to exacerbate flood problems in other areas or reduce the ability of the larger ecosystem to naturally buffer or absorb these natural forces.

There is another important way in which natural disasters and sustainability are related. Most major natural disasters—whether hurricanes, floods, or earthquakes—present clear opportunities to rebuild in ways that promote greater sustainability. By moving people and property out of higher-risk locations, long-term sustainability of a community can be enhanced (for instance, relocating development out of the floodplain as occurred in a significant way following the 1993 Midwest floods). It is important to recognize also that long-term community sustainability can be enhanced in other ways (non-land use), during recovery and reconstruction. For instance, homes and businesses can be rebuilt so that they are more energy efficient, use renewable sources of energy (e.g., solar), minimize waste (e.g., incorporate new water conservation technologies), and minimize the creation of toxins and other harmful pollutants. Increasingly, rebuilding following disasters must be viewed as an important opportunity to move society in the direction of greater sustainability. Often (and perhaps increasingly) these opportunities will be massive. Consider the case of Hurricane Andrew, where some 28,000 homes in Dade County were destroyed, and some 107,000 homes were damaged. With so much housing to replace, even if small improvements were made in the energy and resource efficiency of the new and rebuilt structures, the potential reduction in associated environmental stresses would be considerable.

William Rees has been advocating the use of the concept of the *ecological footprint* as a way to conceptualize and understand the implications of consumption and development patterns, and this concept has tremendous application to natural hazards as well (Rees, 1992; Wackernagel and Rees, 1996). The ecological footprint, simply put, is an estimate of the land and water area needed to support consumption and development practices. Included in these calculations is land necessary to produce food and provide housing, transportation, and other consumer goods. Not surprisingly, Rees concludes that the average North American has a very large ecological footprint (each person requires about five hectares), and that North American lifestyles are supported through the appropriation of carrying capacities of other regions and nations. Ecological footprint analysis can be done at a number of levels (from the individual to the national policy level), and might be adapted to take into account natural hazards and disasters. The costs—environmental and economic—for instance, of locating a housing development in the floodplain might be incorporated into an ecological footprint analysis. The displaced flood waters must be mitigated or managed in some way by society; this might take the form of additional acreage of wetlands downstream or flood storage capacity, or might be expressed in terms of the economic damages from future floods. Also, redevelopment following a disaster, by reducing displacement of floodwaters, for example, or by reducing energy and resource consumption, can serve to substantially reduce the ecological footprint. In these ways, hazard mitigation and sustainability are closely and intimately linked.

What this suggests is that there are several ways of understanding the relationship between sustainability and natural hazards. Exposure to natural hazards clearly reduces a community's sustainability. This vulnerability may be reduced (and the community made more sustainable) through a variety of different measures, including but not limited to land use planning and policy (for instance, by strengthening the community's building code). And natural disasters may present significant opportunities to enhance several different aspects or dimensions of local sustainability, including but not limited to reducing exposure to future disasters. While reducing exposure to natural hazards through land use planning and policy is the primary focus of this book, it is important to keep in mind these other relationships between sustainability and hazards; indeed, the holistic nature of a vision of sustainability requires no less.

Increasingly, hazard mitigation and disaster redevelopment efforts are being described under the rubric of sustainability. A number of recent examples can be cited which illustrate the increasing relevancy of sustainability to natural hazards and which show that these concepts are resonating positively with the natural hazard community, and in particular how land use policies and actions can enhance the sustainability of communities. The granddaddy of these efforts took place in Soldiers Grove, Wisconsin, which chose to relocate its business district entirely out of the floodplain of the Kickapoo River, while at the same time achieving a number of other sustainability objectives, including use of solar energy (new businesses were required to obtain at least half of their

Soldier's Grove, Wisconsin, chose to relocate its business district entirely out of the floodplain of the Kickapoo River, while at the same time achieving a number of other sustainability objectives, including use of solar energy. Like other businesses in the town, Turk's grocery is solar heated (DOE).

energy from solar), conducting life-cycle analysis of building materials, siting buildings and landscaping based on site analyses, as well as mixing housing into the downtown (Becker, 1994). Following the disastrous Midwest floods of 1993, efforts to rethink redevelopment patterns were defined and conceived of in terms of promoting more sustainable communities. A group calling itself the Midwest Working Group on Sustainable Redevelopment formed to encourage flood-damaged communities to think about more sustainable patterns of redevelopment and rebuilding. With U.S. Department of Energy funding, a conference was convened to explore these possibilities, bringing together experts on various aspects of sustainability and community leaders from the region. Several flooded communities took up the charge, in particular Valmeyer, Illinois, and Pattonsburg, Missouri. Valmeyer chose to relocate entirely out the Mississippi floodplain and to create, essentially, a new town designed and built at least in part with sustainability in mind. While not entirely successful, it has encouraged energy conservation and use of solar and geothermal energy. The town of Pattonsburg took a similar approach following the Midwest floods, also choosing to relocate a major portion of the community outside of the floodplain, and again organizing its efforts around the notion of sustainability. The town, in fact, adopted a "Charter of Sustainability" to guide redevelopment, as well as new energy-efficiency and resource-conservation standards, a pedestrian-friendly and solar-oriented new street layout, and a system of constructed wetlands to treat urban run-off, among other features (President's Council on Sustainable Development, 1996, p. 98).

Following the devastating effects of Hurricane Andrew, there were also a number of efforts in south Florida to rethink development patterns more along sustainability principles. Design charettes were conducted, and as a result of one of the more important of these, Habitat for Humanity built an impressive affordable housing project near Homestead. Called "Jordan Commons," the project reflects a new sense among Habitat for Humanity officials that the focus of their efforts should be on creating viable and sustainable neighborhoods, rather than on building individual housing units. The project is unique in its attempt to link safety from future hurricane events (e.g., steel frame structures, which do much better in future events) with a variety of other provisions to support ecological and social sustainability.

As these examples illustrate, some communities are beginning to see reduction of exposure to natural hazards as an essential dimension of their long-term sustainability, and land use planning as an important tool for achieving greater sustainability.

PRINCIPLES OF SUSTAINABILITY/ SUSTAINABLE COMMUNITIES

The broad vision of sustainability which began this chapter was a useful theoretical starting point. For a more detailed vision and more specific guidance regarding the land use policies, programs, and practices described and advocated in this book, an additional set of principles of sustainability or sustainable communities may be useful. Again, I use the term *sustainable communities* (or *community sustainability*, *sustainable community development*, *sustainable cities*) because it orients us toward the local and regional levels, and toward contemplating more sustainable human settlement patterns. The coupling of *communities* with *sustainability* is also important in that it captures the concern about people and overcomes a common criticism that the vision of sustainability is overly or exclusively environmental.

Sustainable Communities Minimize Exposure of People and Property to Natural Disasters; Sustainable Communities are Disaster-Resilient Communities

Any vision or theory of sustainability must prominently include consideration of the long-term safety and survivability of communities and their citizens. Protection from, and avoidance of, natural disasters is an important element of sustainability, and I believe they should receive greater emphasis. A sustainable community, then, is clearly one that seeks to avoid exposure of people and property to natural phenomena like hurricanes, floods, and earthquakes. Communities are *sustainable* when they can survive and prosper in the face of major natural events. Avoidance is the preferable approach, but sustainable communities recognize that some exposure is inevitable (for example, wind forces from hurricanes and coastal storms, which in Florida and other coastal states are unavoidable) and can lead to achieving other important community goals. Minimum building codes and construction standards, for instance, are important components of a sustainable community, as are minimum facilities for sheltering, and programs, policies, and actions for evacuation of residents from high-risk areas.

As the preceding chapters of this book illustrate, there are many different land use tools and techniques promoting avoidance and reducing exposure to natural hazards. They range from land acquisition to land use regulations and to information dissemination. The relocation program undertaken following the 1993 Midwest floods is an example of efforts to avoid disasters, and the actions of communities like Soldiers Grove, Valmeyer, and Pattonsburg, are examples of efforts to create more fundamentally sustainable patterns of human settlements when the long term is considered.

Sustainability does not, however, necessarily mean that all risk is eliminated; sustainable communities strive to balance risk against other social and economic goals. Some degree of risk from natural disasters may be necessary and acceptable. For instance, a number of coastal communities around the country have heavily developed shorelines, subject to extensive natural hazards, including hurricanes, sea-level rise, and long-term erosion. For some communities, strategic retreat or relocation will not be feasible. And, indeed, their location along the beach or shore has economic and aesthetic desirability. The vision of sustainability for communities such as Charleston, South Carolina, or Ocean City, Maryland, will not mean abandonment of their sites, but sensible policies to strengthen buildings, facilitate evacuation, and protect and enhance the natural features and qualities of these places (and, for instance, the choice of beach renourishment over structural hardening of the shore). Sustainability also implies an equitable distribution of the costs of these risks and the programs (such as beach renourishment) for mitigating them.

Where communities choose to allow development to occur in areas exposed to natural hazards, and where there are few alternatives (such as in the case of south Florida), this development can occur in ways that support or advance sustainability, and can generally help create more disaster-resistant communities. Considerable attention in recent years has been focused on development that is sensitive to the environment, and clearly the design of buildings and neighborhoods can reduce long-term risk. Among other things, development can avoid particularly hazardous portions of a site (e.g., occurring away from a surface fault line or away from steep-slope areas), can respect and protect the natural conditions and features that reduce risk (e.g., minimizing impervious surfaces and destruction of vegetation), and can restore and re-create the natural functions that may have been diminished or destroyed (e.g., creation of wetlands, natural drainage swales). (For examples of such environmentally sensitive development projects, see Berke and Beatley, 1992b; Brewster, 1995.) Where possible, ecological infrastructure should take the place of traditional engineering solutions. Such approaches, in working with nature rather than against it, can at the same time accomplish the preservation and enhancement of a range of natural values (e.g., protection of wildlife habitat, creation of significant recreational areas). These projects can often achieve the same results (for example, drainage, flood control) as more conventional engineering solutions, but at significantly lower financial cost.

A sustainable community, then, is a resilient one; it is a community that seeks to understand and live with the physical and environmental forces present at its location.

Sustainable Communities Recognize Fundamental Ecological Limits and Seek to Protect and Enhance the Integrity of Ecosystems

A sustainable community seeks to promote land use and development patterns that acknowledge fundamental ecological limits and operate within the natural carrying capacities of the region. Many of our more serious natural hazards are a direct result of a failure to understand the regional ecological context and to live within it. A sustainable community acknowledges the presence of natural features and processes, such as riverine flooding and natural wildfires, and arranges its land use and settlement patterns so as to sustain rather than interfere with or disrupt them. Riverine systems, for instance, represent areas of biological and ecological importance, and the vision of a sustainable community is one that respects and protects these qualities.

A consistent theme in this book has been the desirability of assuming a regional scale of planning and management. It has been observed in several chapters that the existing system of local governance is fragmented and fractured and does not correspond very well with important natural boundaries. Implementing programs through local government jurisdictions tends to be a fragmented process that works against consideration of the cumulative effects of land use changes over space and time. (How will we know how much, if any, alteration of the watershed can be tolerated before the impacts are unacceptable?) The vision of sustainability, however, strongly supports the need for planning and management at broader regional, bioregional, and ecosystemic levels. Effective strategies for protecting the integrity of the environment, and for addressing a host of important environmental conservation issues (e.g., wetlands protection and restoration, non-point water pollution control, habitat and open space protection) all require broader regional or ecosystem-level strategies.

A sustainability approach to natural hazards understands that frequently the most effective way to reduce vulnerability of people and property is to preserve a healthy, well-functioning ecosystem. There have been considerable experience and consensus over the years concerning the mitigative benefits of the natural environment, for instance the benefits of wetlands, beaches, and dunes (e.g., see Phillippi, 1994-95; Thieler and Bush, 1991). One dramatic case is the decision by the U.S. Army Corps of Engineers to acquire and protect 8,500 acres of natural wetlands in the Charles River (Massachusetts) watershed as an alternative to constructing new flood control works. This strategy of protecting the natural functioning of the environment has proven effective and costs an estimated one-tenth of the cost of flood control works (National Park Service, 1991).

The benefits of a regional, ecosystem approach are perhaps most evident when considering riverine flooding and management of riverine systems. Several earlier chapters have pointed out the destructive effects of upstream actions, such as the building of levees and floodwalls, and the cumulative impacts of discrete land use actions (e.g., the filling of wetlands). The recent report of the Interagency Floodplain Management Review Committee (the "Galloway Report") strongly recommends a move toward more of an ecosystem approach, which acknowledges and protects the natural qualities and functions of the river system (Interagency Floodplain Management Review Committee, 1994).

Successful examples of regional and ecosystem-level management are few. Collaborative management efforts such as the Chesapeake Bay program (and other examples of the Natural Estuaries Program) have had modest, though promising, results (see Chapter 3 for a discussion of this approach). Watershed councils in the northwest and bioregional commissions in California are other promising examples. Effective regional and planning governance is rare in the United States, though the accomplishments of the Portland Metropolitan Service District suggest it is possible. There "Metro" is coordinating an ambitious effort to develop a regional plan for the year 2040. A regional "growth vision" concept has already been adopted, embracing the need to promote compact growth, focus development and redevelopment into existing centers, and protect open space and the natural environment. A more detailed framework plan is being developed to coordinate and guide regional investments and the individual local plans in the region. While Portland's Metro has unique and extensive regional planning powers, it illustrates the importance of planning at this scale for achieving a variety of objectives related to sustainability, including protection of open space and biodiversity, compact development, and a settlement pattern and landscape resilient to natural hazards. As difficult as it may be to bring about, the vision of sustainability is one that assumes, at least in part, an ecosystemic or bioregional management perspective.

Sustainable Communities Promote a Closer Connection With, and Understanding of, the Natural Environment

The vision of sustainability also includes a fuller, deeper understanding of, and connection with, the natural environment. Indeed, there is considerable and increasing evidence of the importance of exposure to nature, especially during childhood, even in very urban environments (e.g., see Kellert, 1996; Nabhan and Trimble, 1994). Sustainable land use patterns attach greater importance to these connections and strive to promote and reinforce them. Floodplains and riparian areas, for instance, begin to be seen as important avenues for people to experience nature and the natural environment. Coastal land use that seeks to direct development back from and away from natural beaches and dune systems acknowledges that doing so preserves these areas for important recreational uses, and can provide opportunities for individuals to experience the coastal natural environment.

Areas subject to natural hazards, whether riverine flooding, wildfires, or seismic mass movement, are seen then as important elements of a land use mosaic that values natural lands as an important part of the human settlement pattern. Even exposure to micro-habitats (i.e., small natural areas) is beneficial, whether it is the nearby butte, the creek behind a row of houses, or the natural space set aside to avoid a surface fault zone. There are increasing examples of environmentally sensitive development projects that, through clustering and creative site design, have been able to set aside such natural areas.

There has been much concern in recent years that our communities and regions do not advance or inspire a sense of place, and that people do not have a sense of being rooted in a particular place. Much of the support for *bioregionalism* is based on developing a greater sense of connection with and commitment to place (e.g., see Sale, 1991). Acknowledging and understanding natural hazards is primary in developing a sense of place. Indeed, the natural forces that shape a place—storms along the coast, movement of the earth, landform, and topography—should be viewed within a sustainability framework as defining elements of a place or region, as fundamental attributes of where we live. Children and adults learn about them, appreciate them, and integrate them into their fundamental understanding of where they live, of their home.

Sustainable Communities Seek to Reduce the Consumption of Land and Resources in Fundamental Ways

At the heart of the idea of sustainability is recognition of *limits*, and of the need to curtail human consumption of land and resources. Without such collective restraint, we will be unable to achieve sustainable communities. The goal of minimizing environmental impacts is certainly not a new one in planning and the other professions with an interest in land. Ian McHarg (1969) made the passionate plea for "designing with nature" more than a quarter of a century ago. Substantial progress has been made in promoting environmental planning and management. Yet a sustainable community is one that does not accept business as usual. It is more than saving a few acres of the habitat of an endangered species, or requiring a few "best management practices" to abate non-point-source pollution. Rather, it represents a basic *shift* in philosophy. Instead of sporadic, incremental adjustments to meet environmental ends, sustainable communities seek to use and consume only the amount of land that is absolutely necessary, and to cause environmental destruction only as a last resort and where no reasonable alternatives exist. The practical implications of this principle for land use are numerous: in essence, land is consumed sparingly, environmentally sensitive land (wetlands, habitat, mountainous areas, coastal shorelines) is placed off-limits, and development and land use practices reduce the consumption of resources at every stage.

This philosophical shift has important implications for the physical form future communities assume. Reducing the inefficient, wasteful, and destructive practice of low-density sprawl becomes a priority in a sustainable community. Cities and towns become more compact and contained, and rely more heavily on infill, redevelopment, and reurbanization within areas already committed to urban development. Compact development patterns can in turn be very complementary to the reduction of natural hazards. Curtailing scattered development and sprawl often means curtailing development in the most dangerous and hazardous locations, as we have seen (e.g., areas prone to wildfires or coastal flooding). It may also allow for more efficient public investments and responses to protect against these hazards (e.g., reducing need for wildfire-fighting capability in certain areas).

Sometimes, however, the goals of compact growth and avoidance of hazardous locations conflict with each other. In some situations containment of growth may actually direct development onto more hazardous sites (e.g., urban floodplains or steep-slope locations previously avoided). Both goals must be considered and high-risk sites within the city avoided—indeed, such areas are important in providing parks and green spaces for urban residents. Recent "reurbanization" efforts by the city of Toronto illustrate how this can occur. Areas of the city have been identified where reurbanization could occur, but the urban floodplain has been designated off-limits. It is possible through such a plan to encourage infill and intensification (reducing the consumption of land at the urban periphery) while at the same time discouraging development of hazardous locations within the city.

As the recent examples of Soldiers Grove, Pattonsburg, and Jordan Commons illustrate, it is also possible to minimize the overall ecological footprint of development at the same time that we achieve safer communities. These examples show that sustainable communities seek to use renewable resources (e.g., solar energy) and limit the consumption of energy and water, at least at the scale of small villages.

Sustainable Communities Recognize the Interconnectedness of Social, Economic, and Environmental Goals

A sustainable community is also one that seeks to integrate social and environmental goals. It is not simply the protection of the natural environment that is important, but the consideration of a broader set of social and economic goals. People are important in this vision, as is the quality of the resulting communities and development. Indeed, the vision of sustainable communities holds that we must consider and promote settlement patterns and policies that accomplish both social and environmental goals.

A number of important social and economic goals are included in the vision of sustainability. These include the availability of affordable housing, transportation, and mobility; access to basic public services and facilities; access to recreation; creation of livable and aesthetically attractive places; and access to jobs, income, and economic activity. The last, economic vitality, is a particularly important component of a sustainable community. To sustain itself in the long run, a community requires the development of a sustainable economic base. In the context of natural hazards, a sustainable economy is one that is resilient and robust in the face of natural events like floods and earthquakes. Keeping businesses and economic infrastructure out of and away from high-risk areas, and areas where natural events cause repeated economic disruption, is one way to go about promoting a more sustainable local economy (along with other mitigation measures such as building codes and construction standards).

The high degree of dependence of many communities on federal and state disaster assistance also raises questions about how sustainable they are. Indeed, the vision of sustainablity in part embodies a spirit of responsibility and self-sufficiency, and heavy reliance on outside resources appears inconsistent with this. Communities must be better prepared to cope with the financial implications of disaster events and should be expected to utilize more of their own resources, at least in all but the most catastrophic of disaster events. Partly this means accepting more responsibility for allowing, or even actively promoting, development in vulnerable places, and striving to reduce this over time; it also means setting aside resources (financial and otherwise) that will be needed to repair and rebuild the community when a disaster occurs.

A sustainable economy includes activities that are environmentally benign and ecologically restorative, which means avoiding economic and industrial activities that heavily degrade the environment, and that destroy or use up the environmental capital of an area. Recall that at the heart of many of the definitions of sustainability is the notion of maintaining ecological *capital*. Increasingly it is recognized that heavily extractive industries (e.g., timber, mining, ranching) do not account for as much economic activity and jobs as once thought, and that maintaining and protecting the natural environment may have a much greater positive economic effect [e.g., creating attractive places to live and retire, or places attractive to new industry—what Power (1996) calls the "environmental view of the economy"].

Protecting the *ecological capital* of a community or region will very often also mean steering development away from high-risk locations and maintaining these areas for their natural qualities, such as coastlines, estuarine shorelines, and areas subject to landslides. In this way, the goals of providing a thriving and healthy economy are inherently interconnected with preserving a healthy environment, and degrading or destroying the latter is not necessary to promote the former. Sustainable communities, then, view the environment, social life, and economy as interconnected. Sustainability is not simply about protecting the natural environment.

Understanding the connections between these different realms is central to the vision of sustainable communities. As Figure 8-1 illustrates, this is the intersection where much of our energy should be focused, and indeed a central goal of sustainability is to increase the overlap between economy, environment, and society; to promote programs, policies, and strategies that simultaneously respond to these different realms (see the report issued by the Governor's Commission for a Sustainable South Florida, as contained in FAU/FIU, 1996). These interconnections have clear relevance to natural hazards. Protecting the integrity and health of ecosystems can at the same time reduce exposure to natural hazards (e.g., by leaving the natural floodplain untouched), enhance the quality of life in the community (by protecting open space, and creating opportunities to enjoy rivers and wildlife), and strengthen the economy and economic base of the community (e.g., by reducing damage and dislocation to businesses).

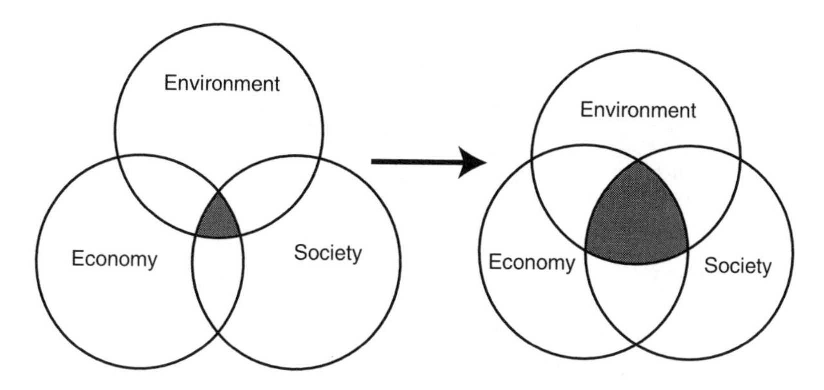

FIGURE 8-1 Sustainable communities work to maximize the overlap between environment, economy, and society. Source: Report of the Governor's Commission for a Sustainable South Florida, as contained in FAU/FIU, 1996.

Sustainable Communities Promote Integrative and Holistic Strategies

Sustainable communities seek to address a number of goals simultaneously and promote strategies and policies that are integrative and holistic. Sustainable communities look for ways of combining policies, programs, and design solutions that bring about multiple objectives. No longer is it possible or desirable, for instance, to view housing policy in isolation from environmental policy or transportation policy or land use policy. This characteristic has clear implications for reducing exposure to natural hazards. Flooding problems in a community may be effectively addressed by certain land use strategies that accomplish other goals at the same time. Certainly protection of wetlands serves to both mitigate flood hazards and to protect important economic and biological resources. Promoting a compact urban form, and urban infill and reuse, can provide more affordable housing at the same time that it prevents or reduces consumption of land at the urban periphery, possibly in turn reducing exposure to natural hazards.

There are many other ways in which natural hazard reduction and other goals can be integrated in the service of community sustainability. The Jordan Commons project mentioned earlier is an example of an effort to simultaneously create a socially cohesive neighborhood, provide safer hurricane-resistant homes, and promote environmental sustainability in the form of more energy- and resource-efficient structures. A number of communities have sought to establish a system of greenways formed in large part from creek beds and floodplains. (These communities include Raleigh and Charlotte, North Carolina; Boulder and Littleton, Colorado; Milwaukee, Wisconsin; Tulsa, Oklahoma; and Scottsdale and Tucson, Arizona. See Little, 1995; National Park Service, 1991; Smith and Hellmund, 1993.) Such greenways protect important ecological features, create areas of special recreational and aesthetic value, and at the same time reduce exposure of people and property to flooding.

Watershed protection efforts, such as the Austin (Texas) Comprehensive Watersheds ordinance, serve simultaneously to protect water quality, reduce flooding, protect important habitat, and set aside valuable open space. Multi-objective river basin planning, which has occurred successfully along several major rivers, is a further example of this integration. Successful efforts to integrate flood-loss reduction, ecological restoration, recreation, and other goals has occurred, for instance, along the Charles River (Massachusetts), the South Platte River (Colorado), the Chattahoochee River (Georgia), and the Kickapoo River (Wisconsin), among others (see National Park Service, 1991). The buyout program following the 1993 Midwest floods has offered similar opportunities to accomplish multiple objectives, and the National Park Service has helped several communities to develop green-space plans for floodplain areas where buyouts and relocation efforts have taken place (Hanson and Lemanski, 1995).

Sustainable Communities Require a New Ethical Posture

Promoting sustainable communities is ultimately about embracing a new ethical posture. There are several important dimensions to this new ethic. Most definitions of sustainability, as we saw earlier, contain at their core a moral reorientation toward the future. The ethic of sustainability explicitly argues for an expansion of the "moral community"—that is, the people or things to which we have a moral duty and whose interests we must consider when making land use and environmental decisions. Under a vision of sustainability it is no longer morally permissible to assume a narrow or traditional short-term time frame in decision making. (This perspective is elaborated on in Beatley, 1994.) The idea of a moral community has significant implications for exposure to natural hazards. Communities have an obligation to consider the impacts of land use policies on future residents and future generations. While a gradual increase in impervious surfaces may have little discernible effect in the short term, what will be the impact on flooding and run-off over a much longer time span? While the construction of a single house on a barrier island may have little effect in the short run, what longer-term development patterns are put into motion by such an approach, and what are the implications for the safety of future residents?

The importance of a longer time frame for mitigation of natural hazards is clear. Much of the current risk and exposure is a result of a past failing to think carefully about the long-term, cumulative effects of individual and community actions—what will be the effect on flooding of continued loss of wetlands, deforestation, and increases in impervious surfaces in the watershed? What will be the effect of allowing incremental development in wildfire-prone areas, or areas of steep slopes? What will be the effect on long-term community risk patterns of building a bridge and causeway system to an undeveloped barrier island? Extending the moral time frame results in greater importance accruing to long-term risk reduction—for example, perhaps imposing a 100- or 200-year coastal setback line, or projecting, say, at least one hundred years into the future the incremental loss of wetlands in a watershed and the likely disastrous results.

The *precautionary principle* often comes up in discussions about sustainability. It states that uncertainty about the extent and nature of environmental impacts should not prevent actions to protect or restore the environment. From an ethical point of view, this principle suggests that we err on the side of caution and conservatism and refrain from actions that may have serious, long-lasting, and potentially irreversible effects, even though such results may not be certain. Under this principle many anthropogenic environmental impacts that we often accept cavalierly would be questionable or unacceptable, such as causing the extinction of a species and loss of biodiversity generally, destruction of wilderness, allowing the continuing emission of greenhouse gases, or the continued loss of topsoil. The precautionary principle suggests that we refrain from altering the natural flow of rivers, that we refrain from structural alteration and reinforcement of the coastline, that we prohibit the filling of wetlands and the destruction of vegetation that provides important wildlife habitat, and that we discourage the needless replacement of natural land with impervious surfaces. Such a principle reinforces the need to take action to prevent exposure of people and property to natural hazards even where their nature and magnitude are uncertain. It supports the need to adjust human settlement patterns to take into account the effects of potential sea-level rise even though there remain uncertainties about its rate and areal effects.

Sustainability also involves an expansion of the moral community spatially and geographically. What obligation does a community have to consider the effects of its actions and policies on other communities and populations, in some cases located many miles away? A community considering, for example, whether to permit development in its floodplain must consider that these decisions may displace floodwaters onto other down-river communities. A community's land use, transportation, and energy policies, as a further example, have significant implications for the generation of CO_2 and other greenhouse gases. The ethical posture of a sustainable community takes these extra-local impacts into account. Sustainability also expands the moral community to consider the interests of other forms of life (see Beatley, 1994, Chapter 7).

The new ethical posture is also one that clearly emphasizes the importance of community, where citizens and community groups are willing and able to see and pursue a larger public good. Self-interest and self-benefit are moderated out of respect for the needs of the larger public, including future generations. At an individual level, a developer understands that his or her desire to build along a high-risk shoreline must be moderated by the larger community's need to preserve the beauty of this land and to ensure the safety of its residents, present and future. The desire of a landowner to erect a new home in the floodplain is restrained by the realization that such an action may have serious ramifications for others and the broader public.

The ethics of sustainability affect how we perceive private property rights. Platt notes in Chapter 2 that recent takings challenges, and the property rights movement generally, are significant impediments to land use planning and management. The new ethical posture would clarify and reestablish the ethical content of private property ownership and use. Private property has never been inviolable and absolute, but rather is subject to collective regulation and control (Beatley, 1994, Chapter 9). The ethics of sustainability suggest that land use is a privilege, collectively bestowed, not an absolute right. No landowners ought to have the right to build houses in the floodway, for instance, if this serves to displace waters onto other people's property or damages the integrity of the riverine ecosystem. No property owners ought to have the unfettered right to, say, fill the wetlands on their land.

In part, the ethical posture toward the use of land and property implied by sustainability ought to be left to individual property owners to translate into action. Indeed, it has proven difficult for government to effectively enforce land use and development regulations (e.g., consider the problems enforcing the south Florida building code, which were evident following Hurricane Andrew). Ideally, the vision of sustainability helps to reinforce a more environmentally and socially responsible ethic of private ownership and use of land. Precisely how this more respectful private land ethic could be nurtured is unclear, but it surely is a function of education, good example, and a system of societal signals and incentives that rewards positive behavior and penalizes negative behavior.

Sustainable Communities Seek a Fair and Equitable Distribution of Resources, Opportunities, and Environmental Risks

Sustainability implies a strong concern for social equity. We must always ask the question, Sustainable for whom? Sustainable communities, then, are places that emphasize social diversity and social, ethnic, and cultural inclusiveness. Sustainable communities are concerned about and strive to promote a fair and equitable distribution of community benefits and opportunities, and a fair sharing of costs and risks.

Natural disasters are not always fair and equal in their impacts, and there are important ways in which a sustainable community in its pursuit of safer, less hazardous land use patterns must be cognizant of social equity. Is risk minimized or reduced for some economic or social classes and not others? Must certain groups endure higher levels of risk? For instance, high-risk floodplains tend to be the location of very inexpensive housing, whose occupants, the community's poorest citizens, are the most vulnerable to flooding. This kind of housing pattern, of course, is not sustainable either from a natural hazard reduction or a social equity point of view, and a program to address the problem might consider these two goals at the same time. A sustainable community seeks to minimize such inequities in response to risks.

Public efforts to make a community more sustainable from a natural hazard perspective may sometimes serve to exacerbate social inequities. For instance, there has been concern that efforts to buy out floodplain properties following the 1993 Midwest floods would have the undesirable effect of reducing the overall supply of affordable housing in riverside communities. As a further example, mandatory seismic retrofit ordinances may have the effect of raising rents and displacing residents; those residents with the fewest economic resources may be the hardest hit by efforts to enhance safety.

In a sustainable community, special efforts are essential to ensure that such effects are minimized. Social equity requires that when a community clears out its floodplain, it must ensure that opportunities for affordable housing are made available elsewhere. And, again, the integrative philosophy implicit in a sustainable community will help. Promoting more compact development, infill, and urbanization, and allowing secondary housing units, for example, can have the important effect of enhancing housing affordability (as in the example of Portland, Oregon).

IMPLEMENTING THE VISION

Building sustainable places means building local capacity to systematically think about, plan for, and effectively shape the future. Several steps in the development process can be identified. An initial step is often understanding the impacts and implications of current patterns and practices of development. A number of analytic techniques and methodologies can prove useful, including the ecological footprint analysis mentioned earlier (e.g., Rees, 1992; Wackernagel and Rees, 1995), carrying capacity studies, and build-out analyses, among others.

The development of sustainability indicators is an important step in understanding the extent to which patterns of development and consumption are or are not sustainable in the long run. A number of communities on the leading edge of the sustainable communities movement have successfully developed such indicators (for example, Seattle, Washington; Santa Monica, California; Cambridge, Massachusetts; Jacksonville, Florida; and the Oregon Benchmarks program; see Community Environmental Council, 1995, for a discussion of sustainability indicators). These indicators, however, typically do not include a natural hazard of disaster vulnerability indicator. Several key indicators could specifically gauge the degree of risk and disaster-resilience of the community, and the extent to which this changes over time. (Are the numbers of people and amounts of property in high-hazard zones on the increase? How many individuals are currently living in seismically deficient structures? What are the trends with respect to impervious surfaces in the watershed?) Sustainability indicators also allow a community to understand its exposure to risk relative to other important community goals, and other resources of sustainability (e.g., availability of open spaces and recreation areas).

Because sustainable community planning is fundamentally participatory and community based, it involves efforts to develop a collective future vision—how the community believes it should evolve and grow over time, and what goals and community attributes people think are important. Very often, design charettes or visioning processes are needed to help in this task. (Such processes have been used very effectively in communities such as Chattanooga, Tennessee; for an extensive discussion of these tools and techniques, see Local Government Commission, 1995.)

Portland, Oregon's 2040 process is exemplary in this regard. Portland Metro has gone to great lengths to engage and consult the public in developing its regional vision, using a variety of methods including a mail questionnaire (with an amazing 17,000+ respondents), community meetings, informational videotapes, and newsletters. Out of this process, a conceptual vision for the future of the region has emerged and will serve as the basis for formulating more specific plans and implementation programs.

Geographic information systems, computer simulations, and other technologies are increasingly useful in helping citizens and community leaders imagine the alternatives available (see also the discussion of GIS in Chapter 5). The California Urban Futures model (CUF) is one recent notable example illustrating the power of these models to show the implications of different community visions, as well as the most effective land use policies and actions to bring them about. The model has been used creatively to show what growth patterns would result in the San Francisco Bay Area under different policy and planning assumptions, including a "business as usual" scenario, a "maximum environmental protection" scenario, and a "compact cities" scenario (see Landis, 1995). This model allows citizens and policymakers to understand the pattern of growth and the consumption of land that would result if current trends are allowed to continue, and if certain sustainable community practices such as growth containment are implemented. The compact cities scenario would result in the development of some 40,000 fewer acres of land than the "business as usual" scenario (a 35 percent reduction in land consumption in the nine-county area, assuming that all new residential density is 18 persons per acre or greater, and that 20 percent of new growth will be accommodated through infill). The power of this technology is that many other "what if" scenarios can be analyzed, depending on the important components of a region's vision. While the "maximum environmental protection" scenario run for the Bay Area included no development on steep slopes or wetlands, it could be expanded, for instance, to more extensively and directly consider natural hazards. What would the regional growth patterns look like if all 100-year floodplains were placed off-limits, as well as high-risk seismic zones (e.g., areas of high liquefaction potential, surface-fault rupture zones, etc.) and other hazardous features and locations? Computer models like the CUF model will be increasingly important in helping to visualize more sustainable places.

A community or region's vision of a sustainable future, including greater resilience to natural forces, should be translated into more concrete "targets" which indicate when and how quickly progress is being made in moving toward sustainability. Hazard reduction targets could be quantitative, and an early draft of the National Mitigation Strategy offers an example of such targets, albeit at the national level. The 1994 draft strategy sets a 25-year goal for mitigation: "By the year 2020, engender a fundamental change in public attitude that demands safer communities in which to live and work, and thereby *reduce by at least half, the loss of life, injuries, economic costs* (1994 dollars), and destruction of natural and cultural resources that result from the occurrence of natural hazards" (Federal Emergency Management Agency, 1994b, emphasis added). Unfortunately, perhaps, the final version of the strategy dropped this quantitative goal. Each locality and region will need to develop its own specific targets and measures of progress, depending upon its specific vision of sustainability.

Once goals and targets are identified, implementation becomes the challenge. There are many opportunities for advancing local and regional sustainability, and many different policy levers and initiatives that might be employed. Many of these have been discussed extensively in the previous chapters. Figure 8-2 presents one initial way to understand these opportunities, which exist at both public and private levels. While much of the emphasis of this book is on public sector planning and policy, the private sector holds much potential to influence and promote safer, more sustainable land use patterns. One important private sector level involves the many decisions that individuals engage in—for instance, making decisions about where to live. Individuals have opportunities to promote sustainable land use by buying homes in safer locations, such as outside of floodplains or off the high-erosion beach front, or away from landslide-prone locations. Such actions reflect the principle that individuals have the responsibility to consider issues of sustainability when making a host of personal choices (which often correspond to self-interest). As Paterson strongly argues in Chapter 7, nongovernmental groups

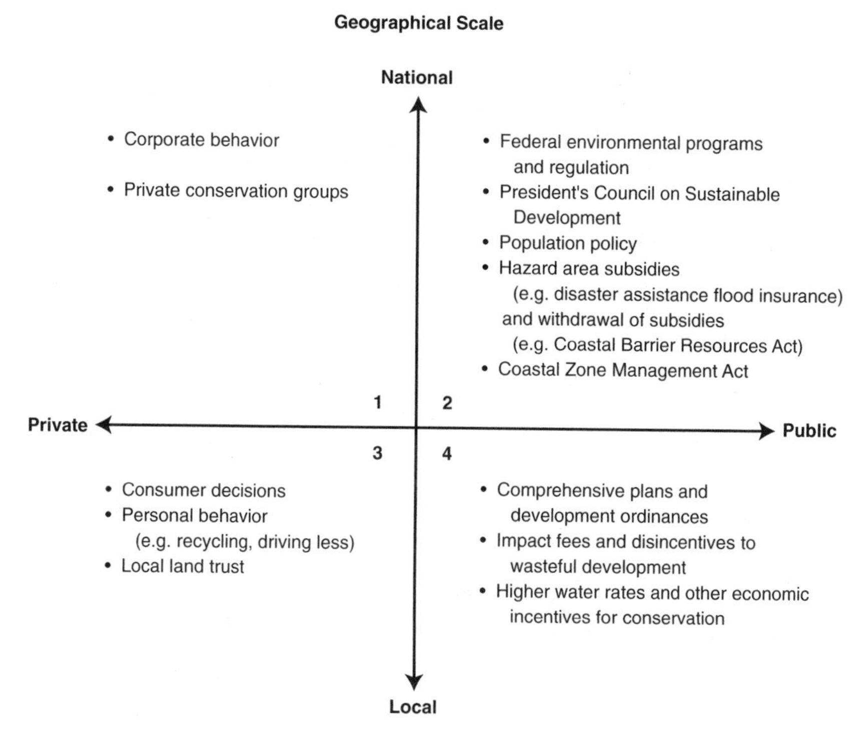

FIGURE 8-2 Opportunities to advance sustainability.

(the "third sector") offer tremendous potential in advancing sustainability. Most of these efforts tend to fall in quadrant three. A host of grassroots environmental groups have a direct interest in hazard mitigation. But environmental groups at the national level, including the National Wildlife Federation and the Coast Alliance, also have been heavily involved in advocating for less destructive development patterns (quadrant 1). Other private sector opportunities include the potentially significant role of groups like the Nature Conservancy, or Habitat for Humanity, or We Shall Rebuild, among many others. These groups can have both direct effects (e.g., buying land, building projects) and more indirect influence in shaping public discourse and dialogue about community sustainability. The sustainability paradigm creates opportunities for different, otherwise disparate, community interests to come together in support of common goals. The emphasis on integrating social and environmental goals means that groups as different as affordable housing advocates, grassroots environmental and open-space conservation groups (e.g., the Greenbelt Alliance, which Paterson cites), taxpayer groups, and members of the business community, among others, can productively join forces and come together in support of a common agenda. Promoting sustainable communities and land use has the potential not only to keep people and property out of harm's way, but also to provide more affordable housing and better living conditions (e.g., higher-density, more compact development, less reliance on the automobile), to protect important environmental resources and amenities, and to reduce the long-term costs of growth and development, among others. The integrative holistic vision of sustainability, then, creates great opportunities for collaboration and political partnerships.

Most of our discussion in this book focuses on public sector land use policy and planning, activities falling on the right side of Figure 8-2. There are a variety of public policy levers and tools that might be effective in promoting more sustainable places. These include land use plans, regulations (e.g., urban growth boundaries to promote compact cities, zoning restrictions to prevent development in high-risk areas), land acquisition, public investment decisions, and taxation and other fiscal incentives.

The other axis of Figure 8-2 reflects the fact that sustainable community policy can be made and actions can be taken at a number of different jurisdictional scales. At the national level, efforts range from federal environmental laws [e.g., the Endangered Species Act (P.L. 93-205, *U.S. Code*, vol. 16, sec. 1531)] and disaster and hazard reduction programs (e.g., the Stafford Act, the National Flood Insurance Program) to the recent activities and recommendations of the President's Council on Sustainable Development.

The sustainability paradigm offers at least the possibility of overcoming the limitations of the present intergovernmental patchwork, described by May and Deyle in Chapter 3. Indeed, the importance of an ecosystem management approach offers the possibility of federal and state agencies coming together to undertake more collaborative, coordinated approaches and decision making. At the federal level, there has been considerable interest in reorganizing federal actions and policy around ecosystem management and achieving the "vertical linkages" Burby discusses in Chapter 1. A federal interagency task force on ecosystem management recently released its report, strongly endorsing the concept of ecosystem management, emphasizing its many benefits, and recommending that federal agencies facilitate the use of ecosystem management (see Interagency Ecosystem Management Task Force, 1995).

Such an approach can be characterized as a "comprehensive regional approach to protecting, restoring, and sustaining our ecological resources and the communities of economies that they support" (p. 17). It would assume, among other things, a broader resource perspective (beyond the narrow focus on one or more resources, such as timber or water), broader geographical and temporal perspectives (overcoming the traditional site-specific focus, and with impacts considered over a much longer time frame). It is also a perspective which is proactive and "aimed at achieving long-term ecosystem conditions, not simply at accommodating short-term demands" (p. 19). Vice President Gore's Performance Evaluation Review report recommends a similar integration of federal policy around ecosystem units. It notes, for instance, that the San Francisco Bay estuary may be affected by some 400 different agencies (National Performance Review, 1993). The sustainability paradigm, in the value it places on ecosystems, has the potential to promote more coordinated collaborative approaches.

Sustainable communities have perhaps the clearest and most potent application on the bottom end of Figure 8-2—at local and regional development levels. Indeed, this is traditionally where the urban and land use policy functions have resided in the United States, and it is at this level where perhaps the greatest potential exists to promote the vision of sustainability.

It is also true, of course, that policies and programs adopted at one scale have profound impacts on sustainability at other scales. Clearly, national policy in a variety of areas influences the local level, and what can and cannot be done there. Federal transportation, housing, environmental, and economic policy, as well as the federal tax code and other de facto policy instruments, have tremendous impact at the local level. Federal programs such as disaster assistance, flood insurance, tax deductions for second homes and casualty losses, and so on, influence the exposure of local land use patterns to natural hazards.

CONCLUSIONS

Sustainability, then, represents a powerful new theoretical framework through which to understand land use and natural hazards. Extremely high levels of property damage in recent disasters clearly suggest that current patterns and practices of land use and community building are not sustainable in the long run. Creating sustainable places means creating places that are far less vulnerable to natural forces and events, and that are resilient to these events. A set of principles has been offered for understanding more specifically what sustainability means at local and regional levels. Among other things, the vision of sustainable communities is one where development and patterns of living acknowledge fundamental ecological and resource limits; seek to reduce or minimize the physical footprint; recognize the interconnectedness of social, economic, and environmental goals; seek integrative and holistic strategies; and acknowledge the need for a new ethic and an equitable distribution of risks and resources. The specific issues, opportunities, and leverage points of a sustainable community, and how to build one, will vary from place to place, but ultimately the sustainable community must derive from an open, participative, and democratic process in which citizens and groups contemplate and coalesce around a desirable view of the future of their community.

CHAPTER NINE

Policies for Sustainable Land Use

RAYMOND J. BURBY[1]

WE CONCLUDE WITH OUR IDEAS about the types of land use policies and the requisite governmental action that could lead to safer and more sustainable communities. For more than two centuries the United States and its people have been on a collision course with nature. Nowhere is this more evident than in the exploitation of shorelines, floodplains, and other areas exposed to natural hazards. The cycle of hazard-zone use, disaster, hazard control, more intensive use, larger disasters, and more elaborate (and expensive) hazard control has been repeated in every state and to varying degrees in most localities. Now hazard policy has reached a breaking point—further large-scale hazard control is unaffordable, the prospect of short-term national economic disruption in the wake of a severe earthquake is no longer just idle speculation, disaster-induced strains on financial institutions such as the U.S. insurance industry are becoming chronic, and the siphoning of billions from the U.S. Treasury for relief is an annual event. To a large extent government policies—federal, state, and local—helped to produce

[1]With contributions from the team of co-authors.

this outcome. In this chapter we show how government policies can help reverse it.

There are several reasons for the seemingly foolish behavior that produces the cycle that proceeds from hazard-zone use to disaster to hazard control and back again. Many hazard zones are attractive for development, for which some level of use is reasonable. The amount of use that makes sense, however, is usually unclear. Risks are poorly understood and frequently misperceived. Even accurate knowledge does not often change behavior, because many people have a high tolerance for risk from natural hazards. The costs of risk avoidance are immediate. The benefits of losses avoided are only apparent at some future date. People tend to greatly discount these benefits, which makes risk avoidance or investment in hazard mitigation unattractive.

Government policies by the score further depreciate the gains from risk avoidance, making those who practice it seem like suckers. Instead, hazard-prone occupants are protected from the full consequences of using these areas by programs to reduce coastal erosion by pumping more sand on ocean beaches (at government expense), to control flooding by building ever higher levees (at government expense), to blunt losses by garnering additional tax write-offs (at government expense), and to shift costs through ever more generous disaster relief payments (at government expense). In short, government policy, rather than countering psychological factors that make it difficult for people to deal rationally with natural hazards, has compounded the problem, making it seem foolish to avoid or limit the use of hazardous areas.

A NEW SET OF PRINCIPLES

The ease of securing government intervention to aid in the use of hazardous areas, or to at least shield users from the adverse consequences of their actions, historically has made restrictive government policy politically unattainable. Nevertheless, we believe change is possible. It can come only if we as a society subscribe to a new set of standards for dealing with natural hazards. The following five principles offer a foundation upon which public policies can be built to break away from the disaster cycle.

- First, governments must limit the practice of subsidizing the risks involved in using hazardous areas.
- Second, governments must build and share a base of knowledge

about the nature of risks and sustainable ways of living with hazards.

- Third, all levels of government must develop commitment and capacity to change the way they manage the use of hazardous areas.
- Fourth, governments must do a better job of coordinating policies to manage exposure to hazards with policies to accomplish economic, social, and environmental objectives through community development.
- Fifth, governments must foster innovation in governance and land management to better match institutional systems and tools with the nature of the problems posed by natural hazards.

In combination, these principles can point the way toward new governmental actions to deal with natural hazards and can be used as a litmus test for evaluating the efficacy of current policies and programs. Gilbert White (1996) has observed that too little attention is paid to post-audits of existing efforts to deal with environmental problems. As a result, little is learned about the effects of public policies in easing (or aggravating) the problems they address, and little progress is made in finding lasting solutions. In the following section, we look at present and past efforts by government to cope with natural hazards to see whether they follow our five principles or have basic flaws that make it impossible to make headway in limiting future disasters. Then we offer our suggestions for new federal and state policies to achieve the vision of sustainability we presented in Chapter 8.

A LEGACY OF MISSED OPPORTUNITIES

Federal and state programs affecting natural hazards have created a legacy of missed opportunities that complicates efforts to find a path to reforms based on appropriate decisions about land use. Political scientist Peter May (1996) notes that this legacy can best be understood in terms of four factors: objectives, approaches, governance, and scope. Looking at objectives, we see how federal and state agencies provide local governments, which bear the brunt of responsibility for land use management, with conflicting (and often confused) signals about hazards. In terms of approaches, we see that federal and state government unwillingness to directly embrace land use management, while pursuing approaches that indirectly skew local land use decisions toward intensive development of hazard zones, effectively limits local discretion to employ the land use approach argued for in this book. Looking at governance, we see the muddle of institutions and competing bureaucracies that disjointed, incremental policy making has created and the need to completely remodel the system now in place. Looking at the scope of hazards policy, we see that most government programs have been narrowly conceived and stand in the way of land use planning and management that is comprehensive in terms of hazards considered, objectives sought, and participants involved in its formulation and execution. In what follows, we explain why the governmental choices of the past have been shortsighted, and therefore are in need of reform.

Objectives

In examining government objectives toward risks posed by natural hazards we find it useful to think first about the types of losses on which government has focused its attention. Policymakers have to deal with four basic kinds of harm: destruction of real and personal property; damage to property; loss of income or increased costs that accrue from losses of property; and injury and death. These losses may occur directly to government personnel and assets, directly to the private sector, and indirectly as with the loss of income to the private sector or tax revenue to government. Once risks are identified, governments have to consider the likelihood that an event will occur and at what severity. In this way, they can differentiate chronic risks (high frequency/low loss) from catastrophic risks (low frequency/high loss) and decide which will be the focus of attention.

In making choices about exposure to loss, government policy has been inconsistent. On the one hand, when government finds it difficult to control the source of risk (that is, to reduce the destructive forces of a hazard), policy tends to focus on limiting the loss of life rather than losses of property or income, and on catastrophic rather than chronic events. Examples of this bias include building codes and emergency warning and evacuation, both of which strongly emphasize life safety rather than protection of property and apply to rarely occurring extreme events, such as hurricanes and earthquakes, rather than chronic events such as flooding. On the other hand, if it is possible technically to control hazards, policy focuses on limiting losses of property, in addition to loss of life, and on chronic rather than catastrophic events. Examples of this bias include policies to deal with flooding through storm drainage and flood control works; shoreline erosion through sea walls, groins, and pumping sand on beaches; and wildfire through enhanced capacity for suppression.

The consequence for localities of these choices made at the federal and state levels are momentous but seldom discussed. Most serious is that while the nation and many states view losses of life in catastrophic events as unacceptable, catastrophic losses of property are viewed as a reasonable risk. The reason is largely economic. Control works capable of stemming truly catastrophic storms are prohibitively expensive and are used only when billions of dollars in property are already at risk. Works that reduce losses in more frequently occurring smaller events (e.g., the 2- to 100-year storm) are more affordable. The consequence, however, is encouragement of development in areas with less than complete protection, and increase in the likelihood that when rarely occurring events happen, they will be accompanied by truly extraordinary losses of lives and property.

In making choices about acceptable risk and creating this consequence, federal and state governments have violated three of the five principles we espouse. First, they have subsidized some (not all) risk by sharing with or wholly relieving local governments and the owners of private property of the cost of hazard control works. Second, they have created a base of knowledge about the location and probability of haz-

Did the Midwest floods of 1993 dampen potential buyers' enthusiasm for this property? (FEMA)

ard occurrence (e.g., seismic, flood hazard and storm surge maps), but the knowledge provided is incomplete in most cases because it does not extend to estimation of losses of lives and property from using hazardous areas. Third, government has not adequately shared decision making nor sought consensus from the whole array of stakeholders. Instead, it has preferred to deal with economic interests and intergovernmental partners and to largely ignore collaboration with other interests (such as good-government and environmental groups) that are less likely to heavily discount future benefits from risk avoidance and might prefer a more cautious stance toward risk.

Government's formulation of objectives for managing risk has long been criticized on other grounds. The most significant criticism is that the federal government and many state governments focus narrowly on risk reduction (e.g., mitigating property and income loss) to the exclusion of the environmental and social equity concerns of people living in and near hazardous areas. Environmental degradation resulting from hazard control measures is too familiar to need elaboration here. Safe development, however, can be equally damaging to the environment when environmentally sensitive areas are cleared, filled, and polluted with stormwater run-off. The neglect of social equity has received less attention, although it too is serious. The narrow focus on property loss reduction benefits property *owners* at the possible expense of *other* community interests who share the cost of local matches for building and maintaining federally subsidized hazard control measures, or who might benefit from dollars spent on other community needs rather than loss reduction. By largely ignoring broader community goals, federal and state policies in dealing with hazards violate our principle of coordinating and integrating policies to accomplish social and environmental, as well as economic, objectives.

Approaches

Choices the federal government and states have made among approaches for dealing with risk are similarly skewed. When faced with risk, policymakers have to decide which of several alternatives will be embraced. They can let people assume risk; or they can try to eliminate or reduce risk; or they can put in place mechanisms to transfer risk. For a variety of reasons, federal and state policymakers have viewed risk assumption and elimination as undesirable and have emphasized risk reduction and risk transfer instead. This bias severely limits choices local governments can make about the use of land in hazardous areas and, in particular, their ability to pursue risk elimination and environmental enhancement as complementary policy objectives. Let us explain.

The loss of life and property chronicled in Chapter 1 long ago led policymakers to conclude that risk assumption—that is, letting individuals and businesses deal with risk, and with damages caused by natural forces, on their own—is unacceptable. But risk assumption may be a rational objective in a number of circumstances if decision makers are given adequate information about the nature of the risks they face. The costs of risk reduction or transfer, for example, may be far greater than losses that are likely to be incurred in a disaster. Individuals and governments have a remarkable ability to absorb loss as a normal operating expense or cost of daily living. One study found, for example, that two-thirds of losses local governments incurred in a typical natural disaster were under $50,000, which is well within the means of all but the most impoverished governments to absorb using available contingency funds (see Burby et al., 1991, p. 31). In spite of the logic of risk assumption and in direct violation of our principles regarding building a shared base of knowledge about risk management, relatively little attention has been given to communicating the risks of natural hazards (beyond short-term warnings to evacuate or take shelter in the face of an oncoming flood, hurricane, or tornado), the importance of responsible individual and local government decision making related to hazards, and appropriate self-protective behavior.

Risk elimination also has received short-shrift when federal and state governments formulate policy, but for opposite reasons. Risk assumption, however rational on economic grounds, is viewed as inviting truly catastrophic losses when individuals and firms expose themselves to risk as a result of misperceiving the potential for loss rather than as a result of rational calculation of gains versus potential losses. Risk elimination—for example, by relocating development out of the way of hazards and preventing future intensive use of hazard zones—would greatly reduce the possibility of catastrophic losses, but at an economic (and consequently political) cost that more often than not is viewed by federal and state policymakers as equally unacceptable.

Risk reduction and risk transfer have been central objectives of federal and state policy toward natural hazards. With the Federal Emergency Management Agency announcement in 1995 that mitigation is now a FEMA priority, risk reduction objectives have assumed even more importance. Risk reduction makes sense when risk lies somewhere in between modest losses best dealt with by risk assumption and catastrophic losses best dealt with by risk elimination (or risk transfer). In this circumstance, employing hazard control measures such as levees or safe development practices such as mandatory building standards may be prudent, at least from the narrow perspective of risk management (more about their environmental costs later).

Risk reduction as a central objective of federal and state policy, however, raises four issues. First, opportunities for cost-effective areawide risk reduction through hazard control are limited at present. Second, while imposition of safe development standards remains a viable option, it suffers from limited applicability to existing development because of economic burdens on property owners, and possibly limited effectiveness for new development because of the potential for widespread evasion by builders and developers. Third, safe development assumes intensive use of hazardous areas is in society's best interest, but these areas often have environmental value that is incompatible with development. Hazard avoidance *plus* environmental gains can justify low-intensity land uses that simultaneously eliminate risk to *both* humans *and* plants and animals. This option is difficult to follow when the federal and state governments single-mindedly promote safe development. Fourth, in some local circumstances it is appropriate to use hazardous areas and risk assumption. When the federal government and states impose uniform risk reduction standards (as with the federal flood insurance program) and uniform treatment of hazards (as in the national model building codes and many state codes), the flexibility to identify areas where risk assumption is appropriate and modify policy objectives accordingly is effectively eliminated.

Transferring risk—through disaster relief, insurance, and tax code provisions that allow deductions for losses in natural disasters—is the other approach favored by the federal government and states in dealing with natural hazards. Since millions of households are at risk in the United States, relief payments and tax deductions help to minimize the human suffering and blunt the adverse financial consequences of disasters. If risk is transferred through insurance, the premiums give decision makers better signals about the true costs of their locational choices. Insurance also has a strong economic justification, since it allows the benefits of using hazardous areas to be realized (in exchange for a small administrative fee and payment of average annual losses), when they otherwise might be foregone because of fears that a disaster would completely wipe out the investment.

To varying degrees, however, both relief and insurance violate our principle of not subsidizing risk. Some economists criticize relief programs for promoting unwarranted and excessive use of hazardous land (most recently, see Lichtenberg, 1994). Insurance is another matter. Initially espoused as a way to share losses *without* subsidizing risk, in actual application it has departed from that principle. In order to garner enough votes for passage, the National Flood Insurance Act of 1968 provided for subsidization of insurance rates for the millions of structures that already existed in flood hazard areas, therefore effectively eliminating relocation (i.e., risk elimination) as a viable policy objective for dealing with those buildings. Twenty-five years after passage of the act, fully 40 percent of flood insurance policies continued to be subsidized (Interagency Floodplain Management Review Committee, 1994, p. 130). A variety of other subsidies then crept into the flood insurance program as administrators found it impossible to adequately adjust premiums upward to account for coastal erosion (National Research Council, 1990), as banks evaded mandatory insurance requirements for mortgages on new structures (U.S. Congress, General Accounting Office, 1990b; this problem was dealt with through new legislation in 1994), and as flood insurance rate maps were not updated to reflect changes in risk.

Subsidies built into federal insurance and relief programs undoubtedly help account for the massive increase in development in hazard-prone coastal regions (Burby and French et al., 1985, p. 88; and Beatley et al., 1994, p. 101). Ironically, the commercial property insurance industry, which helped create this huge increase in exposure to risk by insuring wind damage, has been having second thoughts. It has fostered the creation of special "wind pools" in several hurricane-prone states and is actively lobbying Congress to relieve it of excess risk without, we hasten to add, much attention to the imposition of risk reduction measures to lessen the likely drain on the federal treasury.

In summary, the federal and state focus on risk reduction and risk transfer, besides increasing exposure to risk, has effectively shifted liability for the occupation of hazardous areas to Washington and, to a lesser degree the state capitols, thus relieving local governments of their traditional responsibility for managing these areas and violating our principles regarding risk subsidies and intergovernmental coordination. Federal and state governments have dealt with local stakeholders in a top-down and in some cases highly coercive manner (e.g., the National Flood Insurance Program). Such an approach has done little to foster the "local involvement, responsibility, and accountability" called for in the most recent comprehensive review of federal policy (Interagency Floodplain Management Review Committee, 1994, p. 82) or the spirit of sharing (decisions not just dollars) espoused by the Federal Emergency Management Agency in its 1995 effort to reinvent itself.

In contrast to tendencies on the part of federal and (most) state governments to address hazards in isolation from each other and in isolation from broader societal objectives, more than a dozen states have adopted statewide growth management programs that are much more inclusive. These programs, which we describe in Chapter 3, ask local governments to integrate their methods of dealing with hazards with their methods for reaching an array of other community development goals. State growth management programs use local comprehensive plans as a tool for building knowledge about community problems, building consensus among stakeholders about policy objectives, and for analyzing policy alternatives and deciding upon the best course of action given each community's unique circumstances. In this way, the states encourage localities to pay attention to serious problems, such as natural hazards, but they let each local government craft a solution to the problem that makes sense to it and its citizens. As explained in Chapters 1 and 4, we think this emphasis on planning holds much promise for the nation as a whole. Later in this chapter, we explain how the federal government can foster a planning emphasis in all 50 states.

Governance

The patchwork of governmental institutions and programs recounted in Chapter 3 is a serious obstacle to reform. Governmental leadership from above imposes on localities the policy objectives and approaches favored in Washington and the state capitols, with all of the serious shortcomings noted above. In addition, however, governance is highly fragmented and resistant to change, thus violating our fourth principle, which calls for integration and coordination of efforts to manage exposure to hazards, and our fifth principle, which calls for innovation to craft truly effective measures.

In the case of flooding, for example, 12 federal departments and independent agencies have significant programs dealing with some aspect of the flood problem (Interagency Task Force on Floodplain Management, 1986). Many programs were adopted in the wake of a disaster, but others reflect the entrepreneurial initiatives of individual bureaucrats and departments. The overall result is hazards governance willy-nilly with little consideration for how its various parts fit together across hazards or even among programs dealing with one particular hazard such as flooding or earthquakes.

There have been a number of efforts to rationalize hazards governance at the national level. In 1965 the U.S. Water Resources Council and river basin planning commissions were created to foster interagency and federal-state coordination, but they ceased operation in the early 1980s when funding was eliminated in response to opposition to their command-and-control, top-down style and alleged inefficiencies in their operations (e.g., see Viessman and Welty, 1985). In 1977 President Carter issued Executive Order 11988 directing federal agencies to avoid using floodplains for federal activities and to avoid fostering floodplain development by the private sector. According to the Interagency Floodplain Management Review Committee, however, "it has become apparent that some federal agencies either are unaware of or misunderstand the requirements of the EO and either build or support building in floodplains" so that they "weaken the effectiveness of existing local or state floodplain management regulations and place pressure on local governments to relax their regulations" (1994, p. 78). A similar executive order issued in 1990 (EO 12699) regarding the exposure of federal assets to seismic hazards seems likely to suffer a similar fate.

More recent federal reports that address governance issues, such as the *Unified National Program for Floodplain Management* (Interagency Task Force on Floodplain Management, 1986) or the *Assessment Report* prepared for the same task force (L. R. Johnston and Associates, 1992) have not been acted upon and, at any rate, do little more than describe the irrationality of the system and note the need for better coordination. More visionary efforts, such as the Interagency Floodplain Management Review Committee's *Sharing the Challenge* (1994), have bounced off the same impervious federal bureaucracy and an indifferent Congress with little discernible impact.

The division of responsibility for approaches to hazards is another important aspect of governance. While the federal government has centralized in Washington key aspects of decisions about most ways of dealing with hazards, with only a few exceptions it has steadfastly ignored the planning and management of land use in the mistaken and inconsistent belief that regulation of land is *solely* the responsibility of state and local entities (e.g., see Interagency Floodplain Management Review Committee, 1994, p. 74). Curiously, this belief has not afflicted federal efforts to solve other societal problems. The government directly regulates wetlands, for example. It indirectly regulates, through incentives for state and local planning and regulation, a number of other land use practices to reduce air and water pollution, groundwater contamination, traffic congestion, exposure to airport noise, and coastal zone and other problems. In the case of hazards, however, the government prefers to *act* as if land use is not its concern, while at the same time its policies toward hazards have marked effects on the land use choices available to landowners and local governments.

Just as hazards invariably involve land use and thus require decisions about how land use management responsibilities will be shared, they also tend to produce effects that spill over both state and local governmental boundaries, which requires multijurisdictional cooperation horizontally (among governments at the same level) and vertically (among governments at different levels). This characteristic of hazards has produced a few remarkable innovations in governance (e.g., the Tennessee Valley Authority), but for the most part the federal government and most states have had difficulty formulating and sustaining regional hazard mitigation strategies.

Scope of Policies

Reflecting policy choices made on an ad hoc basis (often in response to needs that became apparent following various natural disasters), federal programs and many state programs for dealing with hazards are partial, not comprehensive. They are partial in that they deal more with some hazards than others (earthquakes, floods, and hurricanes garner most of the attention), and partial in the ways in which they deal with these hazards—that is, in isolation from each other, in isolation from related aspects of disaster management (mitigation, preparedness, response, and recovery), and in isolation from other local, state, and federal policy objectives and programs. While our fourth principle and a number of earlier policy studies and documents call for more comprehensive approaches (as discussed above), the federal government has made few perceptible moves in that direction, although some states have been much more innovative.

The Federal Emergency Management Agency has advocated what it calls an Integrated Emergency Management System (IEMS) (McLoughlin, 1985). However, a similar recognition of the interrelated nature of mitigation, preparedness, response, and recovery and of the need to think in terms of multiple hazards as well as multiple objectives is not evident in other federal agencies. Nor has much progress occurred at the state and local levels, where institutional fragmentation of responsibility for these functions and hazards has so far frustrated efforts by FEMA to take a holistic approach. Typically, at the local level emergency preparedness and response are the domains of civil defense and the fire and police services, while mitigation and recovery are the domains of planning, engineering, and public works departments. Occasionally special agencies, such as flood control districts, have responsibility for particular hazards. The various local departments and agencies with responsibilities related to hazards may be located in separate buildings and in any case communicate with each other infrequently. Similar diffusion of responsibility characterizes most of the states.

More progress has been achieved in integrating hazards considerations into local planning and land use management, particularly in the states that have moved forward with comprehensive growth management programs. Research recounted in Chapter 3 suggests that these programs have fostered attention to hazards in the preparation of local land use plans, which tend to be of higher quality in states with mandates than in states that leave planning and land management solely to local discretion. Plans, in turn, have been found to produce land use management programs that are broader in scope and presumably more effective (Burby and May et al., 1997; May et al., 1996).

A POLICY AGENDA

Viewed against the gloomy picture presented in the preceding pages, it might seem that little progress can be made in fostering the vision of sustainability presented in the preceding chapter. However, we believe progress is possible. Each level of government can take action now to foster sustainability by reducing vulnerability to natural hazards. Although we focus on actions needed from the federal government and states, this does not mean that local governments should stand by idly waiting for initiatives from above. The heart of this book, in fact, is devoted to actions local governments and partners in the private sector can take to foster sustainability. These include systematically assessing vulnerability and risk, establishing planning processes to weigh information and possible courses of action, and working collaboratively with stakeholders to devise land use management strategies. The planning processes we recommend increase the likelihood that local governments will discover ways of dealing with hazards that are feasible given local political, financial, and institutional constraints. In many cases, this planning will reveal that local governments already have in place basic mechanisms such as zoning ordinances and subdivision regulations that can be employed in hazard reduction. The key task, as we explained in Chapters 4 and 8, is to harness the power of planning and land use management to enhance sustainability.

As we see it, the federal government and the states have two important roles to play. The first is to actively support the land use approach by building local commitment to deal with hazards and local capacity to assess vulnerability and risk, plan for sustainability, and manage land use appropriately. The second is to give local land use planning and management a chance by *not* violating the five principles we introduced at the beginning of this chapter and thus skewing local decision making toward unsustainable choices about land use. In the remainder of this section, we examine each of these principles and their implications for federal and state policies.

Our analysis yields an agenda for action and an agenda for research. We call for action because experience with planning for safety from hazards over the past three decades makes some needed actions self-evident, and we can propose policy adjustments with confidence in the results that can be achieved. We call for research because in some areas further research is needed before policymakers can proceed with the ideas we espouse here; these are ideas that have promise, but they need careful study before policymakers can be certain they will be worth the resources required. In addition, research is needed on a number of fronts to further strengthen the capacity of government to deal with hazards.

Government Must Limit the Practice of Subsidizing Risk

The following subsidies seem ripe for careful examination leading to either sharp reduction or outright repeal: disaster relief, flood insurance, shoreline protection, flood control, and tax write-offs of losses to property located in identified hazard zones. In each case, if people are more effectively informed about the risks of natural hazards and if state and local governments adopt appropriate land use management measures, returning risk-management decisions to individuals and businesses will foster more prudent decisions about initial and continued exposure to loss, and it will foster support for local risk reduction efforts.

The Federal Emergency Management Agency and a number of researchers have called for reductions in disaster relief through imposition of thresholds of loss that better reflect the ability to recover from loss without federal subsidies. Like deductibles in most private insurance policies, losses to state and local governments might have to exceed some amount (e.g., $5 per capita) before disaster assistance is provided, and truly extraordinary losses (e.g., over $75 per capita) would be required before the federal government assumes responsibility for all losses. A similar policy makes sense for state governments. This idea could be acted upon without a large investment in additional research.

In 1994 Congress called on the Federal Emergency Management Agency to investigate the impacts of repeal of subsidized flood insurance for structures built prior to the date their communities entered the regular phase of the National Flood Insurance Program. We believe the low rates of voluntary purchase of insurance indicates that subsidies could be repealed with little adverse effect on individuals who own property in flood hazard areas; on the contrary, any adverse effects would be far outweighed by signaling the intent of Congress to return more responsibility for risk management to state and local government and the private sector. Short of outright repeal of these subsidies, we believe they could be reduced if surcharges were placed on premiums charged repetitively flooded structures. These surcharges would give property owners a needed incentive to either relocate buildings beyond the reach of flood waters or retrofit structures to reduce flood damages and the amount of the insurance surcharge. Appropriate adjustments could be made to ensure that surcharges do not work an unjust hardship on low- and moderate-income households.

Congress also has requested study of shoreline erosion rates and the effect of no longer insuring against damage due to erosion. An end to the shoreline erosion subsidy, which is also highly inequitable since it benefits the wealthy much more than others, is long overdue.

The U.S. insurance industry effort in recent years to have the federal government assume greater financial responsibility for property insurance risks from natural hazards is the latest attempt to subsidize risk by shifting costs to the taxpayers. We believe this and all similar efforts to create risk subsidies should be strongly resisted. The private sector must assume responsibility for the consequences of its actions.

To ease the impacts on property owners as subsidies are reduced or eliminated, we believe Congress should appropriate funds to acquire land in hazardous areas for public use (and non-use). This acquisition effort could target repeatedly flooded buildings, which would no longer be eligible for subsidized flood insurance, and property with high environmental value, which will decrease in economic value when it no longer can be developed partially at public expense. In order to reduce the expense of maintaining what would be a widely scattered inventory of property, title to parcels acquired could be transferred to local governments or nongovernmental land trusts such as those examined in Chapter 7.

Finally, we believe federal subsidies for hazard control should be sharply reduced or even eliminated by requiring that most, if not all, of the costs of flood- and other hazard control structures be covered directly by the states, localities, and private interests that would benefit from them. This could be accomplished by phasing out federal appropriations to the agencies providing these services and requiring that they operate on a full cost-recovery basis. States and localities would bear more of these costs initially, but they in turn could pass costs back to the owners and occupants of hazardous areas through mechanisms such as benefit assessment districts and impact fees and taxes. We believe initial steps in this direction can be taken now (for instance, to recover costs from new development in hazardous areas); and, we advocate research to develop cost-recovery mechanisms and estimate likely economic consequences for people and property currently at risk.

Government Must Build a Shared Base of Knowledge

Knowledge about natural hazards and popular understanding of risk has increased enormously over the past three decades, as massive evacuations from hurricane-threatened areas illustrate. However, understanding of long-term, not just immediate, risk is necessary for fully informed decisions about land use. That is a more difficult task, but one that can be achieved.

First and foremost is the need for information about risk, not just about hazards. State and local policymakers and citizens need to know not just that an area is subject to seismic or flood hazards, but that certain levels of loss can be expected with existing or proposed use of such areas, and that those losses will have certain undesirable consequences, particularly if existing federal subsidies are no longer available. For example, if local officials understood that allowing development of flood hazard areas could cause local property tax rates to increase sharply in 1 of every 20 years to cover the loss of tax revenues and infrastructure in the wake of a flood, they might make different decisions when faced with land development proposals. If households purchasing a home in that floodplain had the same information, they might choose a safer location, which would make developers think twice about building in hazardous areas. Currently this information is not available in most jurisdictions, and it is not likely to become available without investment in research to develop and disseminate risk analysis tools to local governments.

In essence what is needed is the type of product-labeling information that has enabled households to make informed decisions about the energy consumption of products they purchase. If hazard-zone property had a similar label, consumers could avoid high-hazard areas, just as they avoid energy-guzzling products, and there would be an incentive to invest in the front-end costs of stronger buildings that would reduce average annual losses.

In addition to information that will facilitate local risk analysis, federal and state agencies can provide information that will encourage local government and citizen use of hazard information. Flood insurance maps are a potentially invaluable source of information about the potential for flooding, but they frequently contain too little additional information (for example, about property lines and existing buildings and landmarks) to be useful in local planning. In addition, detailed mapping needs to be expanded to more communities, and maps need to be more reliable. For land use planning, particularly important improvements would include extension of flood hazard mapping to watersheds of less than one square mile and formulation of floodplain maps that take into account not only existing development but also future development that increases peak discharges from watersheds. The omission of small watersheds in current mapping efforts leaves out the areas of the most rapid urban growth. By adding a full-development scenario to floodplain maps, property owners whose land would become vulnerable to flooding would understand the full consequences of urban growth for their property, and they would likely ask local officials to require detention basins and other measures that reduce peak run-off and the likelihood of flooding. As it stands now, property owners learn the impacts of watershed development after the fact, when what was once flood-free land periodically becomes submerged.

The needed floodplain map improvements have not been implemented for two reasons. First, they will increase the costs of mapping; and, second, they reflect the failure of the flood insurance program to promote the land use adjustments to flooding called for in the National Flood Insurance Act. That is, the program has been operated on the basis of very narrowly conceived insurance aims and not on the basis of the broader loss reduction goals sought by Congress.

Even with their deficiencies, maps provided by the National Flood Insurance Program have enabled far more sophisticated planning for flood hazard areas than was possible prior to the advent of the NFIP. The same can be said for hurricane storm surge maps provided by the government to aid evacuation planning. Local planners would benefit from more widespread federal provision of maps for other hazards, including earthquakes, landslides, and wildfire. Mapping these hazards can be very time- and resource-intensive, as discussed in Chapter 5. But without federal support similar to that provided for the National Flood Insurance Program, needed mapping for hazard delineation is likely to be very slow in coming.

Local decision making can be enhanced by other improvements to the knowledge base. Local hazard assessments cannot be used effectively to support land use planning or decision-making initiatives without empirically validated damage functions, which relate land use and development conditions to probable levels of damage from foreseeable extreme natural events. As we also detailed in Chapter 5, damage functions for private property, public facilities and infrastructure, personal injuries, and death are limited for all hazards except flooding. Better information also is needed on the impacts resulting from lost economic production following natural disasters.

The provision of information and the communication of information are two different matters. In addition to providing better information about hazards, steps need to be taken to ensure that the information is available to and understood by the public. The local planning processes we describe in this book are one important way of doing this, since planning begins by developing information and sharing it with stakeholders. Beyond that, the federal government and states can underwrite public awareness campaigns, as they have for other health risks such as cigarette smoking and auto safety belts. These would include media campaigns, as well as the equivalent of product labeling, such as requiring that information about hazards be included in the deeds to property financed with federal mortgage assistance.

Knowledge must continue to be shared with the various professional groups, in and out of government, whose day-to-day decisions affect exposure to hazards. Targets for improved training materials and courses include professionals in land use planning, civil engineering and public works, building design and code enforcement, public finance, and the emergency services. We think that personnel in each of these professions must have more information in order to see how natural hazards affect their own decisions and how their decisions about hazards affect the choices of sister professions involved in planning and managing urban development. The federal government has taken important steps in working with a number of relevant groups to expand knowledge (e.g., wind and seismic engineering, city and regional planning), which we believe should be continued and expanded.

Finally, knowledge about retrofitting existing buildings to reduce their vulnerability to various hazards has to be diffused among the building trades and do-it-yourself enthusiasts. The national effort to improve the energy efficiency of the building stock can be used as a model. This effort involved all segments of society, from church groups who retrofitted the housing of the poor to utility companies who offered professional assistance with energy audits and rate reductions to households whose homes met energy-efficiency standards. We think that with adequate awareness of hazards and knowledge about how to deal with risk in a cost-effective manner it will be possible to create consensus and commitment to hazard reduction.

Building consensus for action will be more effective if local governments reach out and work with nongovernmental groups. In Chapter 7, we describe a number of groups and ways of working with them to create a broader constituency for efforts to deal with hazards through land use management. Although the federal government and states can work with such groups directly (as they do with the Red Cross, for example), we believe the most effective approach would be to encourage partnerships between local governments and nongovernmental groups to retrofit homes of the poor and elderly located in hazardous areas and to acquire hazardous land as open space and natural areas. These partnerships can be promoted by setting aside grant-in-aid funds for hazard mitigation projects proposed by consortia of governmental and nongovernmental groups, and by authorizing localities to use hazard mitigation grant funds to assist the work of nongovernmental groups.

Government Must Develop Commitment to Manage Hazardous Areas

Lack of state and local government commitment has been the Achilles' heel of previous efforts to deal with natural hazards. Commitment is the willingness of public officials to work energetically to deal with issues posed by natural hazards before, not just after, problems are revealed by a disaster. Previous efforts by the federal government to foster local attention to hazards have produced a sort of pseudo-commitment—officials go through the motions of complying with requirements of the National Flood Insurance Program and preparation of emergency response plans, but their efforts often are careless and only infrequently exceed the bare minimum needed to avoid censure. We think the government can do much more to create genuine commitment, characterized by local enthusiasm for accomplishing sustainable development patterns.

Each of the principles discussed to this point will help build genuine commitment to deal with hazards. A central goal of the planning processes we advocate is social learning, which is why citizen participation is crucial. By increasing personal and community responsibility for the consequences of decisions made, enhancing knowledge, and building understanding of alternative development options, various stakeholders will discover that they can accomplish their goals—economic, social, or environmental—in ways that can be sustained over time. As various groups become mobilized, they will insist that elected officials make decisions about land use that are in the long-term interest of the community.

In the case of flood hazards, the federal government has taken a step in this direction by providing incentives, through the Community Rating System, for local governments to undertake planning processes for flood hazard areas. This is a useful step, but we believe planning must be mandated and not just encouraged through incentives such as insurance rate reductions. The incentive employed by the Community Rating System is effective only *after* local governments experience flood problems because they have allowed extensive development to occur in floodplains. For communities with floodplains that have not been developed extensively (but are highly likely to be in the future), the Community Rating System provides little incentive to begin planning *before* these hazardous areas are committed irreversibly to urban development. A key advantage of mandating planning, rather than just encouraging it, is that communities can find paths to sustainability that are right for their unique local circumstances before (as well as after) development pressures mount.

In addition to mandating planning, the federal government and states can take other actions that will foster genuine commitment. A particular problem has been lagging jurisdictions that for one reason or another fail to embrace sustainability as a growth management goal and ignore natural hazards in making decisions about land development. These governments are untouched by conventional programs that use incentives such as technical and financial assistance to secure local participation; they do not recognize the problems such assistance is aimed at and therefore pass it by. Instead, the federal government and states need to reach out to them with more meaningful measures. In particular, we think that virtually all federal and state aid for infrastructure and economic development could be tied to community participation in land use planning processes, since natural hazards, if ignored, have the potential to wreck these federal investments. Also, recalcitrant local governments could be targeted for particularly close attention in monitoring their compliance with the requirements of a variety of other federal and state assistance programs. The key is to make clear to them that the federal government and states will no longer tolerate inattention to hazards.

Finally, we believe genuine commitment can be enhanced if the insurance industry, once it learns that it cannot foist risk off on the U.S. taxpayer, becomes a strong local advocate for sustainable development. The insurance industry is beginning to take steps in this direction with its initiation of a system to rate local building departments and give insurance rate reductions to property owners in communities with exemplary code enforcement. This will help in developing a constituency for local efforts to reduce susceptibility to hazards. In addition, the industry could enlist its army of local insurance agents as advocates for sustainability. They have a direct economic stake in avoiding disasters and could counteract economic interests that advocate shortsighted development decisions.

Government Must Coordinate and Integrate Hazard Policies

Government efforts to cope with natural hazards are fragmented horizontally at each level of government, vertically between levels of government, and across different types of hazards. This dispersal of authority and responsibility makes it extremely difficult for local governments to deal with hazards in a coherent way. In Chapter 8 we recounted the experience of Soldiers Grove, Wisconsin, in overcoming recurrent flooding from the Kickapoo River. When the village decided to move its central business district out of the river's floodway, it took a flood disaster and the intervention of Wisconsin's two U.S. senators before local officials could sidestep an Army Corps of Engineers proposal to solve flood problems with a dam and levees instead of relocation; then further work was needed to cut through red tape and marshal resources from seven different federal grant-in-aid programs to bring about the relocation. Most localities do not have the time and entrepreneurial talent that it took Soldiers Grove to pursue a land use solution.

The federal government has experimented with a number of different ways of coordinating and integrating hazards policies, but in our opinion none has been particularly effective. These approaches tend to focus on a single hazard, such as earthquakes (e.g., the National Earthquake Hazard Reduction Program) and floods (e.g., the Unified National Program for Floodplain Management). Other federal single-hazard efforts have tried to foster federal-state-local coordination, as in the government's effort from 1965 to 1981 to formulate river basin plans, its attempts to ease hazard mitigation after disasters through the use of interagency hazard mitigation teams, and to prevent future disasters by requiring state hazard mitigation plans. The one federal attempt to integrate policy across hazards, FEMA's Integrated Emergency Management System, has not been replicated by other federal agencies.

Every recent federally inspired evaluation of national hazards policies has noted the problems caused by policy fragmentation and called for more integration. In the latest of these, the Interagency Floodplain Management Review Committee's *Sharing the Challenge* (1994) calls for establishment of an interagency task force to develop a coordination strategy to guide the actions of federal agencies. We believe that, in addition to federal efforts to develop a coordinated strategy, coordination also should occur at the state and local levels, where property is exposed to hazards and the impacts of natural disasters are felt first and most severely. The coordinating mechanism we propose is land use planning. If the needed planning occurs and, just as importantly, federal agencies are required to act in ways consistent with state and local plans, then many of the problems localities experience as a result of fragmentation will disappear.

Government Must Foster Innovation in Governance and Land Management

Innovative governance is the final principle we advocate for guiding reform of federal and state hazard management efforts. Clearly, the context that gave rise to the existing approaches to hazards is changing. The public is disillusioned with large government-funded projects and centralized command-and-control regulation, both of which are viewed as too expensive and inefficient. State and local officials too have increasingly voiced concern about federal mandates that they view as overly expensive, prescriptive, and coercive. Concern for the costs of regulation and government infringement on property rights also is widespread, although in some states and localities support for environmental protection remains strong. This changing context for hazards management calls for rethinking the approaches government uses to bring about sustainable development and reduced risk from natural hazards.

We believe a central role of the federal government must be to foster systematic, collaborative planning processes at the state and local levels rather than to promulgate vast new public investment programs to control hazards or to attempt direct regulation of the use of hazardous areas. Similar to the federal Coastal Zone Management Program's nurturing of state and local planning for balanced use of coastal resources, the new planning processes would encourage all levels of government to define for themselves the meaning of sustainability and the ways in which it can be accomplished. In this way, the federal government would lead by example, providing technical and financial assistance for planning processes states and localities put in place. In place of uniform national goals and standards, federal officials would foster the formulation of goals and standards appropriate to the vastly diverse local situations that exist across America. The federal government's new style would feature cooperation and collaboration with state and local governments, while also being consistent with principles regarding ending subsidies, enhancing information, fostering consensus and commitment, and acting in ways consistent with state and local plans and policies.

THE FEDERAL ROLE IN REALIZING THE PROMISE OF LAND USE PLANNING AND MANAGEMENT

We have shown in this book that land use planning and management can pay large dividends in charting a course toward a safer, sustainable future, and in ensuring that public and private decisions deal with risk and consider community objectives for the use of land exposed to natural hazards. However, as noted in Chapter 6, many local governments, of their own accord, are not likely to employ planning and land use management to the extent that is optimal.

A handful of states have experimented with state growth management programs that feature the formulation of integrated and internally consistent state goals and policies along with state mandates that local governments engage in systematic processes to plan and manage land use. Policy evaluations indicate that these programs can be very effective in dealing with natural hazards and improving sustainability. We believe it is time for the federal government to take steps to speed the adoption of similar programs by all states.

Federal action to foster state and local planning can be thought of in terms of three levels of effort. At the lowest level, incremental adjustments can be made to existing programs. An intermediate level encompasses these incremental changes along with a new executive order to force the new policies upon recalcitrant federal agencies. The third level adds a new, system-changing federal program to bring about responsible land management by state and local governments and the private sector.

Five Changes to Existing Programs

There are five steps the federal government can take *now* to see that state and local planning and land use management take natural hazards into account. First, the Federal Emergency Management Agency could require local preparation of floodplain management plans, already authorized by Congress, as a condition for continued participation in the National Flood Insurance Program. Second, the government could require areawide hazard adjustment plans that give full consideration to land use, not just plans for individual projects, as a condition for federal assistance with hazard management measures such as flood control structures and beach nourishment. Third, the federal government could increase the immediate payoffs for planning by increasing the insurance rate reduction credit given by the National Flood Insurance Program's Community Rating System for floodplain management plans. Fourth, the government could take steps to reduce the costs of planning by expanding hazard-zone mapping efforts and by providing, in addition to maps, risk analyses that help state and local decision makers better understand the trade-offs inherent in developing hazardous areas. Finally, by reducing subsidization of risk, as discussed earlier, the federal government could increase the relevance of state and local planning and land use management to policymakers. We believe each of these actions is justified on the basis of what is already known about land use and hazards and can be mounted without additional research.

A New Executive Order

The small steps that constitute the lowest level of effort will spur needed state and local planning and land use management, but the process could be painfully slow, and a number of federal agencies undoubtedly would never get the message that the land use approach to hazards is both effective and necessary. To combat such foot dragging, a new all-hazards executive order could insist that federal agencies address hazards adequately in their own planning and that they use existing authorities, where appropriate, to foster planning and land use management by state and local governments. An executive order could also specify that federal agencies act in ways consistent with state and local plans, which would also provide an incentive to prepare plans where they do not now exist. This intermediate level of effort also is justified by the existing policy research.

A Federal Hazardous Area Management Act

To have a more dramatic impact, Congress would have to enact new legislation, similar to the Coastal Zone Management Act, to stimulate planning and management of hazardous areas to accomplish a balanced mix of economic development and environmental goals. A national hazardous area management act would not impose national standards on state and local governments, but it would require them to initiate planning and management processes so that exposure to risk from all natural hazards is carefully considered in public and private decisions about development and redevelopment. Obviously, additional research would be required to fully specify the provisions of a new national hazards mandate and to evaluate its likely costs and benefits. Here we provide a brief sketch of the key provisions of the legislation we have in mind.

The key carrot to induce state and local governments to participate could be federal assistance for planning, using some of the funds diverted from disaster relief when subsidization of risk is curtailed. In addition, the government could coordinate and expand its existing hazard-zone mapping and other technical assistance efforts to lower planning costs and provide additional incentives for states and localities to participate. This new national effort could be spearheaded by several agencies, including the Federal Emergency Management Agency (which has hazard mitigation expertise), the Department of Housing and Urban Development (which has community planning and development expertise), or either the Department of the Interior or Department of Commerce (both of which have geophysical and resource management technical expertise).

In fostering state planning processes and state mandates for local planning, the federal government needs to ensure that the states take three additional steps that are essential for success: (1) provide adequate authority for state agencies to monitor and enforce prescriptions about the content of plans (e.g., consistency with state hazard reduction goals) and plan-preparation processes (e.g., adequate risk analysis and citizen participation); (2) take measures to foster commitment to hazard reduction by state and local officials (e.g., adequate sanctions and incentives); and (3) provide tools to build capacity, such as technical and financial assistance, model-plan elements, and hazard maps and other information. In other words, we are calling for plans and land use management efforts that have substance, not paper exercises. Because some localities will lag behind regardless of the strength of state mandates, we also believe it is essential for the states to have in place the ability to regulate hazard-zone development directly, so that future disasters are not created by recalcitrant localities that refuse to adhere to sound planning and land management practices. Many states already have in place critical-area laws of one sort or another (e.g., wetland protection programs) that can serve as models for hazard-zone regulations.

Research Strategies to Sustain Land Use Policy

In addition to charting a new course for federal policy, we believe new strategies are needed for research on natural hazards. Comparison of this book with a similar effort by Baker and McPhee in 1975 indicates dramatic progress in accomplishing four objectives. First, researchers have developed a good understanding of the tools that can be used to manage land use in hazardous areas. Second, a good start has been made in developing the technology and methods needed to identify vulnerability to hazards in ways that are useful for planning. Third, the characteristics of good plans have been identified, and intergovernmental planning processes that can foster such planning have been evaluated. Fourth, evaluations of local government efforts to deal with flood, coastal storm, and earthquake hazards have highlighted factors that foster and hinder land use planning and management. As a whole, this research effort has demonstrated that land use planning and management have promise for reducing losses to life and property and enhancing sustainability, but their potential is not being realized in thousands of hazard-prone communities.

We believe future research should be encouraged in three strategic areas: plan making, vulnerability assessment, and implementation. In the first area, while we know what a good plan looks like, we do not know enough about how to craft such a plan in a way that will foster and sustain commitment to sustainability. In the second, important strides have been made in developing the technology needed to assess vulnerability, but emphasis now needs to be given to translating it into computer software and procedures that are accessible to local government personnel. In the third, we recognize that plans, regulations, and other land use management measures, once adopted, require complex chains of supportive decisions by public agencies and private decision makers if they are to be effective, but the keys to ensuring compliance with planning recommendations and regulatory prescriptions are elusive. These three strategic directions provide guidance to both researchers and research-funding agencies in identifying more specific targets for study. Here we briefly enumerate some of those we think will be most beneficial.

Research on plan making is needed to help planners make selections in each of the four dimensions of choice identified in Chapter 4. Specifically, research should investigate: (1) consensus building and other ways to encourage stakeholders and nongovernmental organizations to participate in plan making, plan adoption, and plan implementation; (2) mechanisms for building the capacity of local and state governments and nongovernmental organizations to analyze hazards and formulate strategies that are cost effective; (3) procedures for making appropriate choices between integrating hazard mitigation into comprehensive plans or making stand-alone plans; (4) factors affecting the flexibility, effectiveness, and equity of various mitigation strategies; and (5) the actual practical value of the principles proposed for creating high-quality plans. We also think research is needed to document the outcomes of hazard mitigation plans in terms of costs incurred by the public and private sectors, gains in reduced vulnerability, and the manner in which planning processes, choices, and principles generate community consensus and effective action. Particularly helpful would be identification of regional approaches to planning and hazard mitigation that have proven to be institutionally feasible and cost effective.

The second strategic direction for future research is vulnerability assessment. New hazard assessment technologies have been developed over the past decade, and some of these systems are widely deployed, but we do not know how (or even if) they are affecting community-wide planning or planning for specific development projects. In a related vein, we need better validation of damage functions across many types of hazards—earthquakes, fire, landslides, storm surges, and wind. With the exception of floods, no consensus has yet developed about appropriate damage functions, which are essential in justifying sustainable land use choices to local officials and, when challenged by aggrieved landowners, in court.

The third strategic direction for future research, and arguably the most important, is investigation of ways to improve the implementation of planning mandates, plans, and hazard mitigation prescriptions contained in building codes and land use regulations. A variety of strategies are available for managing land use, but little is known about their relative effectiveness. Particularly useful would be studies that compare the benefits and costs of strategies that feature avoidance of development in hazardous areas versus those that emphasize safe development practices. Implementation requires commitment to sustainability *and* capacity for undertaking measures that enhance it. Both have been lacking in a significant number of state and local governments and in private sector decision makers, but evidence is accumulating that commitment is the key missing ingredient. Research is needed to find the ways to build commitment that are not overly coercive or costly to undertake. A number of alternative strategies have been identified, some involving the use of persuasive information, some involving incentives, some using monitoring and sanctions to compel compliance, and others using moral persuasion, as argued for in the preceding chapter. We do not know the extent to which these strategies, acting alone or in combination with each other, are effective in building commitment to hazard mitigation and sustainability, but knowing how effective they are is essential for improving the implementation of land use plans and regulatory measures.

While research has established the efficacy of state planning mandates in inducing local planning for hazard mitigation, little is known about other institutional arrangements for land use management. Research is needed to classify and evaluate the success of various institutional arrangements, such as multistate consortia, state seismic safety advisory boards, watershed planning partnerships, and state-local cooperative programs. In addition, research is needed on the roles of key intermediaries and the prospective roles of other entities in translating mandates and plans into action—that is, state and local agencies, nongovernmental organizations, and other interest groups concerned with sustainable futures for communities.

A FINAL NOTE

In conclusion, we have no illusions about the difficulties in bringing about a new land use approach to dealing with natural hazards in thousands of local governments and by millions of public and private decision makers. It will not be an easy task, and it cannot be accomplished by edict. Leadership from the federal government is critical. We call for a new National Hazardous Area Management Act and program to foster improved planning and management at the state and local levels. Short of this step, a new executive order and numerous smaller changes in the way the federal government does business can ensure that its actions more nearly correspond to the principles we espouse. All of these efforts will be more effective if informed by the results of federally funded research to help overcome key stumbling blocks.

State and local governments have a particularly strong interest in and responsibility for prudent planning and management of land exposed to natural hazards. They must collaborate more closely with each other in dealing with exposure to hazards. Collaboration, however, must extend beyond government to embrace professional groups, nongovernmental citizens' groups, and most importantly, private citizens. Critical to all of this is fuller understanding of sustainability so that the concerns about the use of land in hazardous areas expressed in this book are shared widely, and so that consensus begins to form about appropriate courses of public and private action.

APPENDIX

Annotated Bibliography of Selected Research

THIS BIBLIOGRAPHY OF SELECTED research on the effectiveness of local land use planning and management for the mitigation of natural hazards was compiled and annotated by John D. Tallmadge, research assistant at the Department of City and Regional Planning, University of North Carolina at Chapel Hill.

THE FIRST NATIONAL ASSESSMENT

Over 20 years ago, a national assessment of natural hazards research was undertaken to catalogue the status of research and to set funding priorities for the future, and to assess research needs in the area of land use regulations in hazardous areas.

White, Gilbert F., and J. Eugene Haas. 1975. Assessment of Research on Natural Hazards. Cambridge, Mass.: MIT Press.

The authors find that most research had been conducted on the physical and technological aspects of hazards, rather than on the social sciences. Land use management is identified as one of the primary areas of research need common to most hazards.

Baker, Earl and John McPhee. 1975. Land Use Regulations in Hazardous Areas: A Research Assessment. Boulder: Natural Hazards Research and Applications Information Center, University of Colorado.
This assessment includes a discussion of alternate techniques and research needs. The findings include a call for empirical evaluation of whether the goals sought by land use regulation have ever been realized and whether there have been adverse impacts due to existing programs.

MANDATES FOR PLANNING AND LAND USE MANAGEMENT

State mandates for local land use planning has proved to be useful in achieving federal and state goals for the management of land subject to natural hazards.

Mandates and Planning

Burby, Raymond J., and Peter J. May, with Philip R. Berke, Linda C. Dalton, Steven P. French, and Edward J. Kaiser. 1997. Making Governments Plan: State Experiments in Managing Land Use. Baltimore, Md.: Johns Hopkins University Press.
In this book, a team of scholars from six universities uses data collected from more than 150 cities and counties in five states to show that state planning mandates have helped local governments plan for and manage land subject to natural hazards. The authors find that the efficacy of these mandates depends on how well the states craft growth management legislation, how amply programs are funded, and how dedicated state officials are to working with localities. In local areas, they find that success turns on the quality of plans prepared and, as important, the commitment of local officials to state policy objectives. Recommendations are provided to help states craft effective planning mandates.

May, Peter J., Raymond J. Burby, Neil J. Ericksen, John W. Handmer, Jennifer Dixon, Sarah Michaels, and D. Ingle Smith. 1996. Environmental Management and Governance: Intergovernmental Approaches to Hazards and Sustainability. London and New York: Routledge Press.
This book addresses alternative ways in which national or state governments can influence actions taken by local governments to manage hazardous areas. It contrasts the "coercive" intergovernmental approach as found in Florida's growth management program with a "cooperative" intergovernmental approach as found in approaches to hazards management in New Zealand and New South Wales, Australia. The comparisons draw attention to the need for a mix of coercive and facilitative tools in designing

governmental mandates. The book provides detailed descriptions and analysis of the different approaches as carried out in the three settings.

May, Peter J., and Walter Williams. 1986. Disaster Policy Implementation: Managing Programs Under Shared Governance. New York: Plenum Publishing.

This book addresses the "shared governance dilemma" that is brought about by differing incentives for hazard mitigation programs at federal, state, and local levels of government. The authors examine different ways in which federal agencies exert influence over state and local governments, with particular attention to flood management, earthquake preparedness, dam safety, and civil defense planning.

Design of Mandates

May, Peter J. 1993. Mandate design and implementation: Enhancing implementation efforts and shaping regulatory styles. Journal of Policy Analysis and Management 10(2):634–663.

The author examines state planning mandates in five states and concludes that there are several features that can be adapted to influence implementation at the local level. Further, the author concludes that it is easier to influence implementation efforts than regulatory style, and easier to employ formal, legalistic approaches than to foster conciliatory approaches.

Impacts of Mandates on Quality of Local Plans and Development Management

Extensive research has been conducted to specifically explore the relationship between state mandates and the use and effectiveness of land use measures for natural hazard mitigation. The following study found that little hazard mitigation was undertaken in communities in states without such mandates.

Drabek, Thomas E., Alvin H. Mushkatel, and Thomas S. Kilijanek. 1983. Earthquake Mitigation Policy: The Experience of Two States. Boulder: Institute of Behavioral Science, University of Colorado.

The authors performed a case study of earthquake mitigation conditions and strategies in Washington and Missouri through interviews and survey questionnaires with state and local officials, and a review of documents and reports. They found that although there has been historical damage from earthquakes in both states, there has been minimal mitigation activity. Further, it appears that the level of policy activity is related to the frequency of interactions among the stakeholders.

Much of the research on planning mandates has found a strong positive correlation between the existence of a state mandate for including hazard mitigation in land use plans and the quality of local mitigation plans.

Berke, Philip R., Dale Roenigk, Edward J. Kaiser, and Raymond J. Burby. 1996. Enhancing plan quality: Evaluating the role of state planning mandates for natural hazard mitigation. Journal of Environmental Planning and Management 17(2):178–199.

The authors report that state mandates for natural hazards planning result in plans in communities that otherwise would not do planning, and in an improvement in the quality of plans. Where there is no mandate, local plans vary widely, with some communities lagging far behind.

Berke, Philip R., and Steven P. French. 1994. The influence of state planning mandates on local plan quality. Journal of Planning Education and Research 13(4):237–250.

The authors find that state mandates have a positive impact on local hazard mitigation plans. They also find that some mandates are more effective than others. These findings are based on data gathered from 139 communities in five states.

Burby, Raymond J., and Linda C. Dalton. 1994. Plans can matter! The role of land use plans and state planning mandates in limiting development of hazardous areas. Public Administration Review 54(3):229–238.

Based on data from 176 local governments in five states, the authors conclude that land use plans offer communities an opportunity to evaluate the merits and demerits of further development in hazardous areas, leading many communities to subsequent adoption of zoning and other regulations to limit such developments. Also, without state mandates that require inclusion of hazards mitigation in local plans, many local governments will ignore opportunities for risk reduction through planning and development-limiting land use regulations.

Dalton, Linda C., and Raymond J. Burby. 1994. Mandates, plans, and planners: Building local commitment to development management. Journal of American Planning Association 60(Autumn):444–461.

Based on data gathered from local communities in five states, the authors conclude that (1) plans are limited, but important tools in hazard mitigation, (2) local agency commitment and capacity do not vary with the strength of planning mandates, and (3) hazard-specific mandates have affected planning agency commitment to hazard reduction as well as adoption of different approaches to development management.

May, Peter J., and Thomas A. Birkland. 1994. Earthquake risk reduction: An examination of local regulatory efforts. Environmental Management 18(November/December):923–937.

The authors find that local willingness to undertake risk reduction programs is more closely related to local political demands and community resources than to objective risk of previous seismic events. Further, they find that state mandates have had selective effectiveness in achieving compliance with state goals. Findings are based upon data gathered from questionnaires completed by officials from a sample of cities in California and Washington subject to moderate to high seismic risk.

HAZARDS, RISKS, AND LOSS ASSESSMENT

The first step in any planning process for hazard mitigation is to assess the likelihood of a hazardous event and the potential loss if the event were to occur. This is an important step in assessing the need for mitigation measures and for increasing the salience of the issue for interest groups and decision makers.

Burby, Raymond J., with Beverly A. Cigler, Steven P. French, Edward J. Kaiser, Jack Kartez, Dale Roenigk, Dana Weist, and Dale Whittington. 1991. Sharing Environmental Risks: How to Control Governments' Losses in Natural Disasters. Boulder, Colo.: Westview Press.

A comprehensive investigation of the magnitude and character of government losses from natural disasters, a presentation of a range of policy options for addressing those losses, and an evaluation of the opportunities for and constraints on policy innovation and reform. The study includes a chapter on physical planning strategies and a chapter looking at the state of practice based upon a survey of local officials from 481 cities.

Emmi, Philip C., and Carl A. Horton. 1993. A GIS-based assessment of earthquake property damage and casualty risk: Salt Lake County, Utah. Earthquake Spectra 9(1):11–33.

In a study on damage forecasting in Salt Lake County, Utah, the authors employ (1) a microzonation of the earthquake ground shaking hazard; (2) an inventory of buildings by value, structural frame type, and use; (3) earthquake damage functions defining performance of buildings as a function of ground shaking intensity; (4) data on the density of residential and employee populations; and (5) earthquake casualty functions defining casualty risk as a function of building damage.

Office of Policy Development and Research, U.S. Department of Housing and Urban Development. 1995. Preparing for the "Big One": Saving Lives through Earthquake Mitigation in Los Angeles, California. Los Angeles: U.S. Department of Housing and Urban Development.

This report provides estimates of the costs of damage from the Northridge earthquake and the costs of recommended mitigation efforts. Structures considered include schools, health care facilities, residences, and lifelines.

Based on interviews, research reports, and news articles, the authors conclude that an inventory and assessment of local building stock and additional financial resources would be necessary for successful mitigation efforts.

Organization of American States. 1991. Primer on Natural Hazard Management in Integrated Regional Development Planning. Washington, D.C.: Organization of American States.

A primer on the integration of natural hazards assessment and management with regional development planning in Central and South America. The text includes a discussion of the techniques used for natural hazards assessment: remote sensing, Geographic Information Systems, multiple hazard mapping, and critical facilities mapping.

Petak, William J., and Arthur A. Atkisson. 1982. Natural Hazard Risk Assessment and Public Policy: Anticipating the Unexpected. New York: Springer-Verlag.

Using risk analysis techniques, the authors estimate annual expected losses due to natural hazards, identify specific mitigation strategies and technologies and the costs of their implementation, and discuss several policy-making considerations and constraints. From this, the authors develop and assess policy options appropriate for addressing natural hazard risks, and recommend policy actions at the federal, state, local, and private levels.

Spangle, William, and Associates, Inc. 1979. Seismic Safety and Land-use Planning: Selected Examples from California. Washington, D.C.: U.S. Government Printing Office.

This report explains how information on seismic hazards can be effectively integrated into land use planning and decision making to reduce seismic risks. It concludes that land use planning will become a more effective risk reduction strategy as more accurate and detailed information becomes available.

Spangle, William, and Associates, Inc. 1988. Geology and Planning: The Portola Valley Experience. Portola Valley, Calif.: William Spangle and Associates.

The authors explore a case study in the use of geologic information in land use planning. By combining seismic safety goals and avoidance of rupture fault lines and steep slopes with other planning goals, local officials were able to successfully implement several effective techniques.

LAND USE PLANNING APPROACHES AND TECHNIQUES

Much research has been conducted to assess the effectiveness of local hazard mitigation techniques and approaches. This research includes evaluations of both structural and nonstructural (e.g.., land use management) techniques for a variety of natural hazards. Most research has focused on floods, hurricanes, and earthquakes.

One common finding is that pre-disaster planning is, in fact, effective in reducing the loss of life and property in a community. This finding is drawn from research methodologies ranging from case studies of local performance of specific techniques to broad national surveys of the use and effectiveness of mitigation techniques. The research finds that past mitigation efforts have left considerable room for improvement, and recommendations are often made to that end.

Local Case Studies

California Seismic Safety Commission. 1991. California at Risk: Reducing Earthquake Hazards 1992–1996. Sacramento, Calif.: California Seismic Safety Commission's Earthquake Hazard Reduction Program.
Based on lessons learned from the Whittier Narrows and Loma Prieta earthquakes, as well as prior experience, the authors suggest that communities that consider their recovery in advance of a seismic event suffer far less social and economic disruption than communities that do not.

California Seismic Safety Commission. 1995. Northridge Earthquake: Turning Loss to Gain. SSC Report No. 95-01. Sacramento, Calif.: California Seismic Safety Commission.
In this report the authors assess the effectiveness of land use planning in mitigating losses during the Northridge earthquake. Based on background reports, testimony at hearings, issue statements, and 27 case studies of buildings damaged by the earthquake, land use planning is found to have been partially effective, and potentially very effective.

Interagency Floodplain Management Review Committee. 1994. Sharing the Challenge: Floodplain Management in the 21st Century. Washington, D.C.: U.S. Government Printing Office.
This national report makes several findings on the effectiveness of various engineering strategies during the flood of 1993. Recommendations are also made for land use management strategies that would effectively achieve risk reduction, economic efficiency, and environmental enhancement in the floodplain.

Orians, Carlyn E., and Patricia A. Bolton. 1992. Earthquake Mitigation Programs in California, Utah and Washington. Seattle, Wash.: Battelle Human Affairs Research Centers.
This document provides a catalogue of selected seismic hazard mitigation programs in communities in the states of Washington, Utah, and California. The goals and objectives of each strategy are described along with specific mitigation measures. The authors also include other information, where available, such as a brief evaluation of effectiveness, estimated costs of implementation, and procedural and organizational information.

National Studies

Other studies have been broader in scope, assessing the use and effectiveness of specific mitigation techniques for a variety of hazards in communities across the country. The research reveals that land use management does play an important role in reducing the risks to people and property from natural hazards.

Berke, Philip R., and Timothy Beatley. 1992. Planning for Earthquakes: Risk, Politics and Policy. Baltimore, Md.: Johns Hopkins University Press.

Berke, Philip R., and Timothy Beatley. 1992. A national assessment of local earthquake mitigation: Implications for planning and public policy. Earthquake Spectra 8(1):1–15.

Results presented in these two documents were drawn from a nationwide survey of mitigation efforts of local government planning programs and three case studies of such programs. The authors conclude that land use management can play an important role in reducing local seismic risks. They examine a variety of land use management tools and techniques that can be used to reduce seismic risk, and also look at the effects of internal and external factors on the adoption of these techniques. They find that there is much room for adoption of such programs to improve seismic hazard mitigation, particularly through post-earthquake recovery plans.

Burby, Raymond J., Scott A. Bollens, James M. Holway, Edward J. Kaiser, David Mullan, and John R. Sheaffer. 1988. Cities Under Water: A Comparative Evaluation of Ten Cities' Efforts to Manage Floodplain Land Use. Boulder: Institute of Behavioral Sciences, University of Colorado.

The authors assess data from ten cities in terms of their implications for three land use management goals: (1) decreasing future development in the floodplain; (2) increasing the use of protective measures in whatever development occurs in the floodplain; and (3) increasing the proportion of property owners who purchase flood insurance. Research findings were based on comparison of 1985 floodplain conditions with those established in a 1976 study, and on a survey of three groups of decision makers (landowners, developers, and consumers).

Burby, Raymond J., and Steven P. French, with Beverly Cigler, Edward J. Kaiser, David H. Moreau, and Bruce Stiftel. 1985. Floodplain Land-use Management: A National Assessment. Boulder, Colo.: Westview Press.

The authors assess the state-of-practice of floodplain land use management in the United States. Their conclusions are based on two national surveys of localities conducted in 1979 and 1983, and on three case studies. Primary findings include: (1) simple regulations, like those required in the National Flood Insurance Program, are not likely to be effective; (2) the scope of floodplain land use management programs is strongly associated with program effectiveness; (3) land use measures can be effective at preventing flood damage to future development, but not to existing development; and (4) structural measures (especially dikes and levees) can often lead to more development in the floodplain than would otherwise occur.

Godschalk, David R., David J. Brower, and Timothy Beatley. 1989. Catastrophic Coastal Storms: Hazard Mitigation and Development Management. Durham, N.C.: Duke University Press.

The authors find that the single most effective local strategy for hurricane and coastal storm hazard mitigation is to incorporate mitigation objectives into a multiobjective land use management program. The authors also discuss a variety of hazard mitigation policy options with a special emphasis on land use management strategies. The results are based on findings from a national survey of 598 coastal localities and three case studies.

L. R. Johnston and Associates. 1992. Floodplain Management in the United States: An Assessment Report. Volume II: Full Report. Washington, D.C.: U.S. Government Printing Office.

This report is a comprehensive look at floodplain management approaches and techniques used in the United States. Although it found that past floodplain management efforts have accomplished much, losses from flooding continue to rise, and negative effects are seen from traditional structural approaches to modifying flooding. Improvements are suggested to a variety of mitigation strategies, and a coordinated, yet flexible, approach is recommended for a unified national strategy.

Mader, George C., and Martha Blair Tyler. 1993. Land use planning. In Improving Earthquake Mitigation: Report to Congress as Required Under Public Law 101-614 Section 14(b), National Earthquake Hazards Reduction Act. Washington, D.C.: Federal Emergency Management Agency, Office of Earthquakes and Natural Hazards.

An assessment of land use and planning strategies that can be used to reduce earthquake losses. A matrix describing 11 general strategies, the extent of their use, and their general effectiveness is included.

Platt, Rutherford H., H. Crane Miller, Timothy Beatley, Jennifer Melville, and Brenda G. Mathenia. 1992. Coastal Erosion: Has Retreat Sounded? Boulder: Institute of Behavioral Science, University of Colorado.

The authors look at strategies to protect property from the hazard of coastal erosion. They find that the strategies most commonly employed are requirements regarding elevation of new or rebuilt structures and minimum setbacks from shore. Both strategies have shown to have some effectiveness, yet neither strategy is found to be used adequately. The authors offer recommendations for improving federal, state, and local mitigation programs.

Contrary to this line of research, a report was published in 1985 that questions whether in fact it is even possible to measure the effectiveness of hazard mitigation measures.

Natural Hazards Research and Applications Information Center. 1985. Evaluating the Effectiveness of Floodplain Management Techniques and Community Programs. Boulder: Natural Hazards Research and Applications Information Center, University of Colorado.

This report presents the proceedings of a 1985 conference to discuss the state of knowledge about the effectiveness of individual floodplain management techniques and community programs. This conference found that for a number of reasons it is very difficult, if not impossible, to determine the extent to which a particular technique or program will minimize losses and prove cost-effective.

Hazard-Specific Measures

Other research has found that the effectiveness of a specific mitigation technique is often affected by the type of hazard. Depending upon local conditions and the hazard to be addressed, different combinations of techniques often demonstrate greater effectiveness than techniques used singly.

Reitherman, Robert. 1992. The effectiveness of fault zone regulations in California. Earthquake Spectra 8(1):57–77.

The subject of this paper is the effectiveness of the Alquist-Priolo program in California and its applicability to other states and hazards. The author finds that land use planning and engineering techniques are effective in combination but demand differing emphases depending upon the seismic hazard (e.g., surface rupture versus liquefaction versus ground shaking).

Spangle, William, and Associates, Inc. 1989. The Post-Earthquake Hazard Mitigation Process: A Revision and Recommendation Following the Whittier Narrows Earthquakes of 1987. Portola Valley, Calif.: William Spangle and Associates.

This report, based on interviews with local officials and a review of reports on the Whittier Narrows and the Alaska earthquakes, determined that the type of damage sustained during an earthquake affects the strategies considered for post-earthquake response. The Alaskan earthquake, which resulted in ground failure, led local officials to consider amending building codes and rezoning hazardous land.

Sustainability

Recently, researchers have begun to evaluate the effectiveness of hazard mitigation measures by a new yardstick: sustainability. The following work begins to probe how this might change our view of past and current management practices and offers suggestions for future innovation.

Beatley, Timothy. 1995. Planning and sustainability: The elements of a new (improved?) paradigm? Journal of Planning Literature 9(May):383–395.

Beatley, Timothy. 1994. Promoting sustainable land use: Mitigating natural hazards through land use planning. In Natural Disasters: Local and Global Perspectives. Boston, Mass.: Insurance Institute for Property Loss Reduction.

These two articles summarize literature on the state of the art of land use and development planning techniques to mitigate hazards and promote sustainable land use patterns. The author recommends incentives to localities to explore how new planning tools might work. Such tools include: transfer of development rights, clustering, traditional neighborhood development ordinances, greenway systems, innovative acquisition efforts, promotion of more compact and contiguous development patterns, hazard mapping, emphasis on multiple objectives, and linking hazard mitigation to the sustainable communities movement.

FACTORS INFLUENCING EFFECTIVENESS

Research has shown that a number of factors, in addition to state planning mandates, often influence the effectiveness of hazard mitigation measures. One such factor is the extent to which the mitigation approach is linked with other community goals or with the mitigation of other hazards. This often happens through the normal land use planning process.

Comerio, Mary C. 1992. Impacts of the Los Angeles retrofit ordinance on residential buildings. Earthquake Spectra 8(February):79–94.

The author evaluates the Los Angeles Earthquake Hazards Reduction Ordinance of 1981. Upon examination of the compliance rate, costs of implementation, and effects on rents for residential buildings, she suggests that seismic safety policy must be linked with affordable housing goals. If this strategy is to succeed in other cities, innovative financing strategies must be considered for retrofit ordinances of unreinforced masonry buildings.

Quarantelli, Enrico L. 1991. Disaster response: Generic or agent-specific? In Managing Natural Disasters and the Environment, Alcira Kreimer and Mohan Munasinghe, eds. Washington, D.C.: World Bank.

The author finds that although the technical and physical aspects of disaster mitigation planning are agent-specific (i.e., particular to the type or class of disaster), the human, group, organizational, community, and social aspects tend to be generic. The author argues that a generic, or all-hazards, approach to disaster planning would be more effective than a hazard-by-hazard approach.

Wyner, Alan J., and Dean E. Mann. 1986. Preparing for California's Earthquakes: Local Government and Seismic Safety. Berkeley, Calif.: Institute for Governmental Studies, University of California.

The authors of this research report use case studies to understand how selected California communities have planned and implemented seismic hazard mitigation strategies. They find that use of land use measures in the Seismic Safety Elements (SSEs) of local comprehensive plans in California has become routine for many communities. Seismic hazards are considered when evaluating a development proposal, yet rarely do such concerns prevent approval of the development.

Much research has been conducted to examine the political, economic, and technological conditions and processes that influence the effectiveness of local mitigation strategies.

Alesch, Daniel J., and William J. Petak. 1986. The Politics and Economics of Earthquake Mitigation. Boulder: Institute of Behavioral Science, University of Colorado.

The authors conducted case studies and surveys of three California cities (Long Beach, Los Angeles, and Santa Ana) to examine the development, enactment, implementation, and effects of a code to reduce the number of unreinforced masonry buildings. Their findings relate to political, economic, and technological conditions that contribute to effective use of seismic safety techniques.

Berke, Philip R. 1989. Hurricane vertical shelter policy: The experience of Florida and Texas. Coastal Management 17(3):193–217.

The author examines case studies of Texas and Florida to explore the political conditions under which communities consider using vertical shelters (multistory buildings structurally reinforced to withstand hurricane forces) as an emergency evacuation option. He finds that certain conditions can be promoted to enhance the likelihood of adoption and the effectiveness of implementation.

Burby, Raymond J., and Steven P. French. 1981. Coping with floods: The land-use management paradox. Journal of the American Planning Association 47(3):289–300.

The authors identify a paradox in floodplain land use management: the same local conditions that might stimulate adoption of such programs also serve to create development pressures in the floodplain, thus compromising the effectiveness of land use strategies. Based on a survey of 1,203 local jurisdictions, they find that land use measures are less effective when there is little vacant land available for future development. Thus if land use measures are to be effective, they must be put in place well before pressures for floodplain development begin to mount.

Rossi, Peter H., James D. Wright, and Eleanor Weber-Burdin. 1982. Natural Hazards and Public Choice: The State and Local Politics of Hazard Mitigation. New York: Academic Press.
The authors use survey techniques to examine the salience of natural hazard problems to elites, the assessment of risks from hazards among the larger population, and the patterns of elite and group activity in hazard mitigation policy formulation. They found that natural hazard mitigation is of low political salience to the general population and to elites, and that very little opinion is held about the various mitigation options.

Bibliography

Aberly, Douglas, ed. 1994. Futures by Design: The Practice of Ecological Planning. Philadelphia, Pa.: New Society Publishers.

Abt, Robert, David Kelly, and Mike Kuypers. 1987. The Florida Palm Coast fire incidence and residence characteristics. Fire Technology 23(3):186–197.

Adler, Robert, Jessica Landman, and Diane Cameron. 1993. The Clean Water Act 20 Years Later. Washington, D.C.: Island Press.

Advisory Committee on the International Decade for Natural Hazard Reduction. 1987. Confronting Natural Disasters: An International Decade for Natural Hazard Reduction. Washington, D.C.: National Academy Press.

Alesch, Daniel J., and William J. Petak. 1986. The Politics and Economics of Earthquake Hazard Mitigation. Boulder: Institute of Behavioral Science, University of Colorado.

Alfors, John T., John L. Burnett, and Thomas E. Gay, Jr. 1973. Urban Geology Master Plan for California. Bulletin 198. Sacramento, Calif.: California Division of Mines and Geology.

Algermissen, S.T., and Karl V. Steinbrugge. 1978. Earthquake losses to buildings in the San Francisco Bay area.

In Proceedings of the Second International Conference on Microzonation for Safer Construction—Research and Application. San Francisco: The Conference.

All-Industry Research Advisory Council. 1989. Surviving the Storm: Building Codes, Compliance and the Mitigation of Hurricane Damage. Chicago: The Council.

American Institute of Architects. 1991. Regional and Urban Design Committee: Component Handbook for Disaster Assistance Programs. Washington, D.C.: AIA.

American Institute of Architects. 1996. Buildings at Risk: Seismic Design Basics for Practicing Architects. Washington, D.C.: AIA.

Applied Technology Council. 1985. ATC-13: Earthquake Damage Evaluation Data for California. Redwood City, Calif.: The Council.

Applied Technology Council. 1991. ATC-25: Seismic Vulnerability and Impact of Disruption on Lifelines in the Coterminous United States. Redwood City, Calif.: The Council.

Association of Bay Area Governments. 1995. Shaken Awake! Preliminary Estimates of Uninhabitable Dwelling Units and Peak Shelter Populations in Future Earthquakes Affecting the San Francisco Bay Region. Oakland, Calif.: The Association.

Baker, Earl, and John McPhee. 1975. Land Use Regulations in Hazardous Areas: A Research Assessment. Boulder: Natural Hazards Research and Applications Information Center, University of Colorado.

Balsillie, James. 1983. On Determination of When Waves Break in Shallow Water. Beaches and Shores Technical and Design Memorandum No. 83-3. Tallahassee, Fla.: Florida Department of Natural Resources, Division of Beaches and Shores.

Bay Area Regional Earthquake Preparedness Project (BAREPP). no date. Living on the Fault. (Public information brochure.) Oakland, Calif.: BAREPP.

Bay Area Regional Earthquake Preparedness Project (BAREPP). 1990. Probabilities of Large Earthquakes in the San Francisco Bay Region, California. USGS Circular 1053. Washington, D.C.: U.S. Government Printing Office.

Beatley, Timothy. 1994. Ethical Land Use: Principles of Policy and Planning. Baltimore, Md.: Johns Hopkins University Press.

Beatley, Timothy. 1995a. Mitigating natural hazards. Urban Land 81(October): 214–219.

Beatley, Timothy. 1995b. Planning and sustainability: The elements of a new (improved?) paradigm? Journal of Planning Literature 9(May):383–395.

Beatley, Timothy, and David J. Brower. 1986. Public perceptions of hurricane hazards: Examining the differential effects of Hurricane Diana. Coastal Zone Management Journal (14):3.

Beatley, Timothy, and David J. Brower. 1993. Sustainability meets Main Street: Principles to live and plan by. Planning 59:74–78.

Beatley, Timothy, David J. Brower, and Anna K. Schwab. 1994. An Introduction to Coastal Zone Management. Washington, D.C.: Island Press.

Bebbington, Anthony. 1993. Governments, NGOs, and agricultural development: Perspectives on changing interorganizational relationships. Journal of Development Studies 29(2):199–221.

Becker, William J. 1994. The case for sustainable redevelopment. Environment and Development (November):1–4.

Bell, W. G. [1920] 1971. The Great Fire Of London in 1666. Westport, Conn.: Greenwood Press.

Bender, Stephen O. 1995. Protected areas as a protection against natural hazards. In Expanding Partnerships in Conservation, Jeffrey A. McNeely, ed. Washington, D.C.: Island Press.

Bennett, Jon. 1995. Meeting Needs: NGO Coordination in Practice. London: Earthscan Publications.

Berke, Philip R. 1989. Hurricane vertical shelter policy: The experience of Florida and Texas. Coastal Management (17):193–217.

Berke, Philip R. 1995. Natural hazard reduction and sustainable development: A global assessment. Journal of Planning Literature 9(4):370–382.

Berke, Philip R., and Timothy Beatley. 1992a. A national assessment of local earthquake mitigation: Implications for planning and public policy. Earthquake Spectra 8(1):1–15.

Berke, Philip R., and Timothy Beatley. 1992b. Planning for Earthquakes: Risk, Politics and Policy. Baltimore, Md.: Johns Hopkins University Press.

Berke, Phillip R., and Timothy Beatley. 1995. After the Hurricane: Linking Recovery to Sustainable Development in the Caribbean. Unpublished Manuscript. Chapel Hill : Department of City and Regional Planning, University of North Carolina at Chapel Hill.

Berke, Philip R., and Steven French. 1994. The influence of state planning mandates on local plan quality. Journal of Planning Education and Research 13, 4 (Autumn):237–250.

Berke, Philip R., and Maria Manta. 1997. Sustainable Development and Comprehensive Plans: The State of Practice. Cambridge, Mass.: Lincoln Institute of Land Policy.

Berke, Philip R., and Carlton Ruch. 1985. Application of a computer system for hurricane emergency response and land use planning. Journal of Environmental Management (21):117–134.

Berke, Philip R., and Suzanne Wilhite. 1988. Local Mitigation Planning Response to Earthquake Hazards: Results of a National Survey. College Station, Tex.: Center for Hazard Reduction and Community Rehabilitation, Texas A&M University.

Berke, Philip R., Timothy Beatley, and Suzanne Wilhite. 1989. Influences on local adoption of planning measures for earthquake hazard mitigation. International Journal of Mass Emergencies and Disasters (7):33–56.

Berke, Philip R., Jack Kartez, and Dennis Wenger. 1993. Recovery after disaster: Achieving sustainable development, mitigation, and equity. Disasters 17(2):178–199.

Berke, Philip R., Dale J. Roenigk, Edward J. Kaiser, and Raymond J. Burby. 1996a. Enhancing plan quality: Evaluating the role of state planning mandates for natural hazard mitigation. Journal of Environmental Planning and Management 37(2):155–169.

Berke, Philip R., Jennifer Dixon, Neil Ericksen, and Jan Crawford. 1996b. What Makes a Good Plan? Principles and Methods for Evaluating Plans. Hamilton, New Zealand: Centre for Environmental and Resource Studies, University of Waikato.

Bernknopf, Richard L., Russell H. Campbell, David S. Brookshire, and Carl D. Shapiro. 1988. A probabilistic approach to landslide hazard mapping in Cincinnati, Ohio, with applications for economic evaluation. Bulletin of the Association of Engineering Geologists 25(1):39–56.

Bernstein, George. 1993. Insurance issues in hazard mitigation. In Improving Earthquake Mitigation: Report to Congress. Washington, D.C.: Federal Emergency Management Agency.

Bingham, Richard D., Brett W. Hawkins, John P. Frendreis, and Mary P. Le Blanc. 1981. Professional Associations and Municipal Innovation. Madison: University of Wisconsin Press.

Blowers, Andrew, ed. 1993. Planning for a Sustainable Environment. London: Earthscan Publications.

Bolin, Robert. 1993a. Household and Community Recovery after Earthquakes. Boulder: Institute of Behavioral Science, University of Colorado.

Bolin, Robert. 1993b. The Loma Prieta earthquake: An overview. In The Loma Prieta Earthquake: Studies of Short-Term Impact, Robert Bolin, ed. Boulder: Institute of Behavioral Science, University of Colorado.

Bollens, Scott A. 1993. Restructuring land use governance. Journal of Planning Literature 7(3):211–226.

Bolton, Patricia A., and Carlyn E. Orians. 1992. Earthquake Mitigation in the Bay Area. BHARC-800/92/041. Seattle, Wash.: Battelle Human Affairs Research Center.

Bolton, Patricia A., Susan G. Heikkala, Marjorie M. Greene, and Peter J. May. 1986. Land-use Planning for Earthquake Hazard Mitigation: A Handbook for Planners. Boulder: Natural Hazards Research and Applications Information Center, University of Colorado, Special Publication 14.

Boore, David M., and William B. Joyner. 1994. Prediction of ground motion in North America. In Proceedings of ATC-35 Seminar on New Developments in Earthquake Ground Motions Estimation and Implications for Engineering Design Practice. Redwood City, Calif.: Applied Technology Council, ATC-351.

Borcherdt, Roger D. 1994. New developments in estimating site effects on ground motion. In Proceedings of ATC-35 Seminar on New Developments in Earthquake Ground Motions Estimation and Implications for Engineering Design Practice. Redwood City, Calif.: Applied Technology Council, ATC 35-1.

Boston Globe. March 5, 1992. Private rights, public benefits. 14.

Boswell, Michael R., Robert E. Deyle, Richard A. Smith, and E. Jay Baker. 1997. A Quantitative Method for Estimating Public Costs of Hurricanes. Tallahassee, Fla.: Florida Planning Laboratory, Department of Urban and Regional Planning, Florida State University.

Boulder County Land Use Department. 1994. Boulder County Site Plan Review, Wildfire Mitigation Plan. Boulder, Colo.: The Department.

Boulding, Elise. 1991. The old and the new transnationalism: An evolutionary perspective. Human Relations 44(8):789–805.

Brabb, Earl E. 1984a. Innovative approaches to landslide hazard and risk mapping. In IV International Symposium on Landslides, volume 1. Toronto, Canada.

Brabb, Earl E. 1984b. Minimum Landslide Damage in the United States, 1973–1983. Open-File Report 84–486. Reston, Va.: U.S. Geological Survey.

Brabb, Earl E., and Betty L. Harrold, eds. 1989. Landslides: Extent and economic significance. Proceedings of the 28th International Geological Congress: Symposium on Landslides, Washington, D.C., 17 July 1989. Brookfield, Vt.: A. A. Balkema.

Brewster, George. 1995. A better way to build. Urban Land 54(6):30–35.

Bronson, William. 1986. The Earth Shook; The Sky Burned. San Francisco: Chronicle Books.

Brown, David L. 1991. Bridging organizations and sustainable development. Human Relations 44(8):807–831.

Brown, Lester, et al. 1995. State of the World, 1995. Washington, D.C.: Worldwatch Institute.

Building Performance Assessment Team. 1992. Preliminary Report in Response to Hurricane Andrew, Dade County, Florida. Washington, D.C.: Federal Emergency Management Agency.

Building Seismic Safety Council. 1991. NEHRP Recommended Provisions for the Development of Seismic Regulations for New Buildings, volumes 1 and 2. FEMA No. 222, Earthquake Hazards Reduction Series 16. Washington, D.C.: Federal Emergency Management Agency.

Building Seismic Safety Council. 1992. NEHRP Handbook for the Seismic Evaluation of Existing Buildings. FEMA No. 178, Earthquake Hazards Reduction Series 47. Washington, D.C.: Federal Emergency Management Agency.

Burby, Raymond J. 1995. Coercive versus cooperative pollution control: Comparative study of state programs to reduce erosion and sedimentation pollution in urban areas. Environmental Management 19(3):359–370.

Burby, Raymond J. 1996. Is federal disaster relief at odds with mitigation? Forum for Applied Research and Public Policy 11(3):109–113.

Burby, Raymond J., and Linda C. Dalton. 1994. Plans can matter! The role of land use plans and state planning mandates in limiting the development of hazardous areas. Public Administration Review 54(May/June):229–238.

Burby, Raymond J., and Steven P. French. 1981. Coping with floods: The land use management paradox. Journal of the American Planning Association 47(July):289–300.

Burby, Raymond J., and Robert G. Paterson. 1993. Improving compliance with state environmental regulations. Journal of Policy Analysis and Management 12(Fall):753–772.

Burby, Raymond J., and Steven P. French, with Beverly Cigler, Edward J. Kaiser, David H. Moreau, and Bruce Stiftel. 1985. Flood Plain Land Use Management: A National Assessment. Boulder, Colo.: Westview Press.

Burby, Raymond J., and Peter J. May, with Philip R. Berke, Linda C. Dalton, Steven P. French, and Edward J. Kaiser. 1997. Making Governments Plan: State Experiments in Managing Land Use. Baltimore, Md.: Johns Hopkins University Press.

Burby, Raymond J., Scott A. Bollens, James M. Holway, Edward J. Kaiser, David Mullan, and John R. Sheaffer. 1988. Cities Under Water: A Comparative Evaluation of Ten Cities' Efforts to Manage Floodplain Land Use. Boulder: Institute of Behavioral Science, University of Colorado.

Burby, Raymond J., Beverly A. Cigler, Steven P. French, Edward J. Kaiser, Jack Kartez, Dale Roenigk, Dana Weist, and Dale Whittington. 1991. Sharing Environmental Risks: How to Control Governments' Losses in Natural Disasters. Boulder, Colo.: Westview Press.

Burton, Ian, Robert W. Kates, and Gilbert F. White. 1993. The Environment as Hazard, 2nd ed. New York: The Guilford Press.

California Governor's Office of Planning and Research. 1990. State of California General Plan Guidelines. Sacramento, Calif.: The Office.

California Seismic Safety Commission. 1995. Northridge Earthquake: Turning Loss into Gain. Report No. 95–01. Sacramento, Calif.: The Commission.

Calthorpe, Peter. 1993. The Next American Metropolis: Ecology, Community and the American Dream. Princeton. N.J.: Princeton Architectural Press.

Campbell, Charles O. 1988. Government relations. In Principles of Association Management: A Professional's Handbook. Washington, D.C.: American Society of Association Executives.

Carnegie Commission. 1993. Facing Toward Governments: Nongovernmental Organizations and Scientific and Technical Advice. New York: The Carnegie Corporation.

Carroll, Thomas. 1992. Intermediary NGOs: The Supporting Links in Grassroots Development. West Hartford, Conn.: Kumarian Press.

Central United States Earthquake Preparedness Project. 1990. Estimated Future

Earthquake Losses for St. Louis City and County, Missouri. FEMA No. 192, Earthquake Hazards Reduction Series 53. Washington, D.C.: Federal Emergency Management Agency.

Chiu, T.Y., and Robert G. Dean. 1984. Methodology on Coastal Construction Control Line Establishment. Beaches and Shores Technical and Design Memorandum No. 84–6. Tallahassee, Fla.: Florida Department of Natural Resources, Division of Beaches and Shores.

Cigler, Beverly A., and Raymond J. Burby. 1990. Local flood hazard management: Lessons from national research. In Cities and Disaster, Richard T. Sylves and W.L. Waugh, eds. Springfield, Ill.: Charles C Thomas Publishing.

Clarke, Caroline L., and Mohan Munasinghe. 1995. Economic aspects of disasters and sustainable development: An introduction. In Disaster Prevention for Sustainable Development: Economic and Policy Issues, Mohan Munasinghe and Caroline Clarke, eds. Washington, D.C.: The International Bank for Reconstruction and Development/The World Bank.

Clifton, Robert L., and Alan M. Dahms. 1993. Grassroots Organizations, 2nd ed. Prospect Heights, Ill.: Waveland Press.

Coastal Planning and Engineering/URS Consultants. 1992. Coastal Structure Damage Methodology. Final Report. Paramus, N.J.: Coastal Planning and Engineering.

Cohrssen, John J., and Vincent T. Covello. 1989. Risk Analysis: A Guide to Principles and Methods for Analyzing Health and Environmental Risks. Washington, D.C.: Council on Environmental Quality.

Colorado Office of Emergency Management. 1993. Colorado Natural Hazards Mitigation Council, Annual Report. Golden, Colo.: Department of Local Affairs.

Colorado Office of Emergency Management. 1994. Colorado Natural Hazards Mitigation Council, Annual Report. Golden, Colo.: Department of Local Affairs.

Community Environmental Council. 1995. Sustainable Community Indicators: Guideposts for Local Planning. Santa Barbara, Calif.: CEC.

Congressional Research Service. 1992. FEMA and the Disaster Fund. Washington, D.C.: CRS.

Crowell, Mark, Stephen P. Letherman, and Michael S. Buckley. 1993. Shoreline change rate analysis: long-term versus short-term data. Shore and Beach 61(2):13–20.

Culliton, Thomas J. 1991. Fifty Years of Population Growth Along the Nation's Coast 1960–2010. Washington, D C: National Oceanic and Atmospheric Administration, U.S. Department of Commerce.

CUSEC Journal. 1996. Earthquake loss estimation: The basis for risk reduction planning. 3(1):3–5.

Dacy, Douglas C., and Howard Kunreuther. 1969. The Economics of Natural Disasters. New York: The Free Press.

Dalton, Linda C., and Raymond J. Burby. 1994. Mandates, plans and planners: Building local commitment to development management. Journal of the American Planning Association 60(Autumn):444–462.

David, Elizabeth, and Judith Mayer. 1984. Comparing costs of alternative flood hazard mitigation plans. Journal of the American Planning Association 50(Winter):22–35.

Davis, James B. 1988. The wildland-urban interface: What it is, where it is, and its fire management problems. In Protecting People and Homes From Wildfire in the Interior West: Proceedings of the Symposium and Workshop. General Technical Report-251. Dry Branch, Ga.: United States Forest Service, Southeastern Forest Experiment Center.

Davis, James F., John H. Bennett, Glenn A. Borchardt, James E. Kahle, Salem J.Rice, and Michael A. Silva. 1982a. Earthquake Planning Scenario for a Magnitude 8.3 Earthquake on the San Andreas Fault in Southern California. CDMG Special Publication 60. Sacramento, Calif.: California Division of Mines and Geology.

Davis, James F., John H. Bennett, Glenn A. Borchardt, James E. Kahle, Salem J. Rice, and Michael A. Silva. 1982b. Earthquake Planning Scenario for a Magnitude 8.3 Earthquake on the San Andreas Fault in the San Francisco Bay Area. CDMG Special Publication 61. Sacramento, Calif.: California Division of Mines and Geology.

Debo, Thomas N., and Andrew J. Reese. 1995. Municipal Storm Sewer Management. Boca Raton, Fla.: Lewis.

DeGrove, John. 1992. Planning and Growth Management in the States. Cambridge, Mass.: Lincoln Institute of Land Policy.

Department of Public Works, Government of New South Wales. 1986. Floodplain Development Manual, PWD 860.10, December. Sydney, Australia: The Department.

Department of Public Works, Government of New South Wales. 1990. Coastal Management Manual, September. Sydney, Australia: The Department.

Deutsch, Kenneth. 1996. American Red Cross Mitigation Paper. Washington, D.C.: American Red Cross.

Deyle, Robert E. 1995. Integrated water management: contending with garbage can decisionmaking in organized anarchies. Water Resources Bulletin 31(3):387–398.

Deyle, Robert E., and Richard A. Smith. 1994. Storm Hazard Mitigation and Post-Storm Redevelopment Policies. Tallahassee, Fla.: The Florida Planning Laboratory, Department of Urban and Regional Planning, Florida State University.

Diamond, Henry L., and Patrick F. Noonan, eds. 1996. Land Use in America. Washington, D.C.: Island Press.

DiMaggio, Paul J., and Helmut K. Anheier. 1990. The sociology of nonprofit organizations and sectors. Annual Review of Sociology (16):137–59.

Drabek, Aron G. 1987. Development alternatives: The challenge for NGOs—An overview of issues. World Development 15(Supplement):ix–xv.

Drabek, Thomas E. 1983. Human System Responses to Disaster: An Inventory of Sociological Findings. New York: Springer-Verlag.

Drabek, Thomas E. 1990. Emergency Management: Strategies for Maintaining Organizational Integrity. New York: Springer Verlag.

Drabek, Thomas E., Alvin H. Mushkatel, and Thomas S. Kilijanek. 1983. Earthquake Mitigation Policy: The Experience of Two States. Boulder: Institute of Behavioral Science, University of Colorado.

Dunham, Allison. 1959. Flood control via the police power. University of Pennsylvania Law Review (107):1098–1131

Durham, Thomas. 1993. The role of regional consortia in hazard vulnerability reduction and preparedness. Natural Hazards Observer 18(1):1–2.

Dynes, Russell R. 1993a. Disaster reduction: The importance of adequate assumptions about social organization. Sociological Spectrum (13):175–192.

Dynes, Russell R. 1993b. Social science research: Relevance for policy and practice. In Improving Earthquake Mitigation: Report to Congress. Washington, D.C.: Federal Emergency Management Agency.

Dzurik, Andrew, Bruce Stiftel, Anita Tallarico, and Angel Cardec. 1990. Probable Property Damages from Hurricanes on Florida's Barrier Islands: Development of Method and Application to Gasparilla Island. Tallahassee, Fla.: Environmental Hazards Center, Institute for Science and Public Affairs, Florida State University.

Earthquake Engineering Research Institute. 1994. Expected Seismic Performance of Buildings. Oakland, Calif.: Ad Hoc Committee on Seismic Performance, EERI.

Einsweiler, Robert. 1975. Comparative description of selected municipal growth guidance systems: A preliminary report. In Management and Control of Growth: Issues, Techniques, Problems, Trends, volume 2, Randall Scott, David J. Brower, and Dallas Miner, eds. Washington, D.C.: Urban Land Institute.

Eisner, Richard K. 1991. Regional disaster preparedness: Lessons from the October 17th, 1989 Loma Prieta earthquake. In Proceedings of the 3rd United States/Japan Workshop on Urban Earthquake Hazard Reduction. Honolulu, Hawaii. November 13–15, 1991.

Elfring, Chris. 1989. Preserving land through local land trusts. BioScience 39(2):71–74.

Ender, Richard L., and John Choon K. Kim, with Lidia L. Selkregg and Stephan F. Johnson. 1988. The design and implementation of disaster mitigation policy. In Managing Disasters: Strategies and Policy Perspectives, Louise K. Comfort, ed. Durham, N.C.: Duke University Press.

England-Joseph, Judy A. 1995. Disaster Assistance: Information on Expenditures and Proposals to Improve Effectiveness and Reduce Future Costs. GAO/T-RCED-95-140. Washington, D.C.: General Acounting Office, U.S. Congress.

Estes, Carroll L., Elizabeth A. Binney, and Linda A. Bergthold. 1989. How the legitimacy of the sector has eroded. In The Future of the Nonprofit Sector: Challenges, Changes and Policy Considerations, V. A. Hodgkinson and R. W. Lyman, eds. San Francisco: Jossey-Bass.

FAU/FIU Joint Center for Environmental and Urban Problems. 1996. Excerpts from the Initial Report of the Governor's Commission for a Sustainable South Florida. Environmental and Urban Affairs Winter:19–30.

Federal Emergency Management Agency, Federal Insurance Administration. 1981. National Flood Insurance Program Flood Insurance Manual Revision: V Zone Rates and Rules. Washington, D.C.: FEMA.

Federal Emergency Management Agency. 1987. Final Report on the Evaluation of Three Earthquake Education Projects. Washington, D.C.: FEMA, Office of Natural and Technological Hazards Programs.

Federal Emergency Management Agency. 1990. Post-Disaster Report for Loma Prieta Earthquake. Washington, D.C.: FEMA.

Federal Emergency Management Agency. 1991. The Loma Prieta Earthquake: Emergency Response and Stabilization Study. Washington, D.C.: U.S. Government Printing Office.

Federal Emergency Management Agency. 1992a. Building Performance: Hurricane Andrew in Florida. Washington, D.C.: FEMA, Federal Insurance Administration, December 21.

Federal Emergency Management Agency. 1992b. Federal Response Plan. Washington, D.C.: FEMA.

Federal Emergency Management Agency. 1992c. Interagency Hazard Mitigation Team Report in Response to the August 24, 1992 Disaster Declaration for the State of Florida. FEMA-955-DR-FL Hurricane Andrew. Washington, D.C.: FEMA.

Federal Emergency Management Agency. 1993. Building Performance: Hurricane Iniki in Hawaii: Observations, Recommendations, and Technical Guidance. Washington, D.C.: FEMA.

Federal Emergency Management Agency. 1994a. A Multi-Objective Planning Process for Mitigating Natural Hazards. Denver, Colo.: FEMA, Region VIII.

Federal Emergency Management Agency. 1994b. Natural Mitigation Strategy. Draft. Washington, D.C.: FEMA.

Federal Emergency Management Agency. 1994c. Preserving Resources Through

Earthquake Mitigation. NEHRP Biennial Report to Congress, FY 93–94. December. Washington, D.C.: FEMA.

Federal Emergency Management Agency. 1994d. Strategic Plan: Partnership for a Safer Future. December. Washington, D.C.: FEMA.

Federal Emergency Management Agency. 1995a. Flood Insurance Rate Review - 1995, Depth Percent Damage - Non-Velocity Zones. Washington, D.C.: FEMA.

Federal Emergency Management Agency. 1995b. Part One - Fact Sheet on Reinventing FEMA, National Performance Review, Phase II. Washington, D.C.: FEMA, March 27.

Federal Emergency Management Agency. 1996. Community Rating System Coordinator's Manual. Washington, D.C.: FEMA.

Federal Interagency Floodplain Management Task Force. 1992. Floodplain Management in the United States: An Assessment Report, volume 1. Summary Report. Washington, D.C.: FEMA.

Ferretti, Joan, Donald W. Stever, and Eliza A. Dolin. 1995. Wetland protection. In Law of Environmental Protection, volume 2, Sheldon M. Novick et al., eds. Deerfield, Ill.: Clark Boardman and Callaghan.

Ferris, James M., and Elizabeth Graddy. 1989. The fading distinctions among the nonprofit, government, and for-profits. In The Future of the Nonprofit Sector: Challenges, Changes and Policy Considerations, V. A. Hodgkinson and R. W. Lyman, eds. San Francisco: Jossey-Bass.

Finney, Mark A. 1996. FARSITE: Fire Area Simulator, User and Technical Documentation. Missoula, Mont.: Systems for Environmental Management.

Foster, Harold D. 1980. Disaster Planning: The Preservation of Loss and Property. New York: Springer-Verlag.

Foundation Center. 1995. National Guide to Funding for Community Development. New York: The Foundation.

Fox, Thomas H. 1987. NGOs from the United States. World Development 15(supplement):11–19.

Freilich, Richard, and R. Doyle. 1994. Takings legislation: misguided and dangerous. Land Use Law and Zoning Digest 46(October):3–5.

French, Steven P. 1990. A preliminary assessment of damage to urban infrastructure. In The Loma Prieta Earthquake: Studies of Short-Term Impact, Robert Bolin, ed. Boulder: Institute of Behavioral Science, University of Colorado.

French, Steven P., and Mark S. Isaacson. 1984. Applying earthquake risk analysis techniques to land use planning. Journal of the American Planning Association 50(4):509–522.

French, Steven P., and Lyna L. Wiggins. 1989. Computer adoption and use in California planning agencies: Implications for education. Journal of Planning Education and Research 8(2):97–107.

French, Steven P., Xudong Jia, Elizabeth F. Meyer, and Sanjay Grover. 1996.

Estimating Societal Impacts of Infrastructure Damage with GIS. Buffalo, N.Y.: National Center for Earthquake Engineering Research, State University of New York at Buffalo.

Friedman, Don G. 1974. Computer Simulation in Natural Hazard Assessment. Hartford, Conn.: The Travelers Insurance Company.

Friedman, Don G. 1975. Computer Simulation in Natural Hazard Assessment. Boulder: Institute of Behavioral Science, University of Colorado.

Friedman, Don G., and T.S. Roy. 1966. Simulation of Total Flood Loss Experience of Dwellings on Inland and Coastal Flood Plains. Hartford, Conn.: The Travelers Insurance Company.

Friedmann, John. 1992. Empowerment: The Politics of Alternative Development. Cambridge, Mass.: Blackwell Publishing.

Friesma, H. Paul, James Caporaso, Gerald Goldstein, Robert Lineberry, and Richard McCreary. 1979. Aftermath: Communities After Natural Disasters. Beverly Hills, Calif.: Sage Publications.

Fuller, Margaret. 1991. Forest Fires: An Introduction to Wildland Fire Behavior, Management, Firefighting, and Prevention. New York: John Wiley and Sons.

Futrell, William. 1986. Private sector initiatives in shaping floodplain and wetlands policies. In Proceedings from Floodplain Management Conference, Gainesville, Fla.: The Conference.

Geis, Donald, and Tammy Kutzmark. 1995. Developing sustainable communities: The future is now. Public Management (August):4–13.

Godschalk, David R. 1987. The 1982 Coastal Barrier Resources Act: A new federal policy tack. In Cities on the Beach, Rutherford Platt, ed. Chicago: University of Chicago Press.

Godschalk, David R. 1992. Implementing coastal zone management 1972–1990. Coastal Management 20(3):93–116.

Godschalk, David R. 1995. The record of mitigation under the Stafford Act since 1988. Paper presented at the National Mitigation Conference, Washington, D.C., December 7.

Godschalk, David R. 1996. Assessing Planning and Implementation of Hazard Mitigation Under the Stafford Act. Working Paper #1. Chapel Hill: Center for Urban and Regional Studies, University of North Carolina at Chapel Hill.

Godschalk, David R., David J. Brower, and Timothy Beatley. 1989. Catastrophic Coastal Storms: Hazard Mitigation and Development Management. Durham, N.C.: Duke University Press.

Godschalk, David R., David W. Parham, Douglas R. Porter, William R. Potapchuk, and Steven W. Schukraft. 1994. Pulling Together: A Planning and Development Consensus-Building Manual. Washington, D.C.: Urban Land Institute.

Gori, Paula I. 1991. Communication between scientists and practitioners: The important link in knowledge utilization. Earthquake Spectra 7(1):89–95.

Green, T. 1994. First Impressions of the 1994 Chelan County Fires, About Growth. Seattle, Wash.: Washington State Community, Trade, and Economic Development Agency.

Greenbelt Alliance. 1996. Fact Sheet. San Francisco: The Alliance.

Greenwood, David J., and Darryl J. Hatheway. 1996. Assessing Opal's impact. Civil Engineering 66(1):40–43.

Gurin, Maurice, and Jon Van Til. 1989. Understanding Philanthropy: Fundraising in perspective. In Compendium of Resources for Teaching about the Nonprofit Sector, Voluntarism and Philanthropy. Washington, D.C.: Independent Sector.

Haas, J. Eugene, Robert W. Kates, and Martyn J. Bowden, eds. 1977. Reconstruction Following Disaster. Cambridge, Mass.: MIT Press.

Habitat for Humanity. Undated. Jordan Commons: A Homestead Habitat for Humanity Neighborhood. Homestead, Fla.: Habitat for Humanity.

Haeberle, S. H. 1989. Planting the Grassroots: Structuring Citizen Participation. New York: Praeger Press.

Hall, Peter D. 1995. Theories and institutions. Nonprofit and Voluntary Sector Quarterly 24(1):5–13.

Hamilton, Robert. 1992. In planning for sustainable development and environmental protection, don't overlook the threat of natural hazards. Natural Hazards Observer 16(6)4–5.

Hanks, Thomas, and Hiroo Kanamori. 1979. A moment magnitude scale. Journal of Geophysical Research (84):2348–2350.

Hansen, Andrew. 1984. Landslide hazard analysis. In Slope Stability, D. Brunsden and D.B. Prior, eds. New York: John Wiley and Sons.

Hansmann, H. 1989. The two nonprofit sectors: Fee for service verses donative organizations. In The Future of the Nonprofit Sector: Challenges, Changes and Policy Considerations, V. A. Hodgkinson and R. W. Lyman, eds. San Francisco: Jossey-Bass.

Hanson, Kate, and Ursula Lemanski. 1995. Hard-earned lessons from the Midwest Floods: Floodplain open space makes economic sense. River Voices 6(Spring)1:16–17.

Hardin, Garrett. 1968. The tragedy of the commons. Science (162):1243–1248.

Hart, Earl. 1994. Fault-Rupture Hazard Zones in California. Special Publication 42. Sacramento, Calif.: California Division of Mines and Geology.

Hart, Gary C. 1976. Natural Hazards: Tornado, Hurricane, Severe Wind Loss Models. Report No. NSF/PRA–7509998/4. Redondo Beach, Calif.: J.H. Wiggins.

Havlik, Spencer W., and B. Dorsey. 1993. To plan or not to plan: Opportunities for natural hazards education in graduate planning programs. Natural Hazards Observer 18(2):4–5.

Hay, Claire M. 1994. A brief description of the Boulder County, Colorado, Wildfire Hazard Mitigation System (WHIMS) Hazard Risk Rating Model. Longmont, Colo.: Environmental Modeling and Spatial Analysis.

Hebert, Paul J., Jerry D. Jarrell, and Max Mayfield. 1993. The Deadliest, Costliest, and Most Intense United States Hurricanes of this Century (and Other Frequently Requested Hurricane Facts). Washington, D.C.: National Technical Information Service.

Hocker, Jean W. 1996. Patience, problem solving, and private initiative: Local groups chart a new course for land conservation. In Land Use in America, Henry L. Diamond and Patrick F. Noonan, eds. Washington, D.C.: Island Press.

Hodgkinson, V. A., and R. W. Lyman. 1989. Key challenges facing the nonprofit sector. In The Future of the Nonprofit Sector: Challenges, Changes and Policy Considerations, V. A. Hodgkinson and R. W. Lyman, eds. San Francisco: Jossey-Bass.

Hofner, J. 1990. Status and relevance of NGOs in view of mounting international interdependence. International Association 42(1):4.

Holway, James M., and Raymond J. Burby. 1993. Reducing flood losses through local planning and land use controls. Journal of the American Planning Association 59(Spring):205–216.

Hoyt, William G., and Walter B. Langbein. 1955. Floods. Princeton, N.J.: Princeton University Press.

Hutton, Janice R., Dennis S. Mileti, William B. Lord, John H. Sorensen, and M. Waterstone. 1979. Analysis of Adoption and Implementation of Community Land-use Regulations for Floodplains. San Francisco: Woodward-Clyde Consultants.

Hwang, H., and Huijie Lin. 1993. GIS Database for Mapping Seismic Hazards and Seismic Risk in Memphis and Shelby County, Tennessee. Memphis, Tenn.: Center for Earthquake Research and Information, Memphis State University.

Innes, Judith Eleanor. 1992. Group processes and the social construction of growth management: Florida, Vermont, and New Jersey. Journal of the American Planning Association 58(4):440–453.

Insurance Institute for Property Loss Reduction. 1995a. Coastal Exposure and Community Protection: Hurricane Andrew's Legacy. Boston: The Institute.

Insurance Institute for Property Loss Reduction. 1995b. Increase in coastal exposures highlights need for hazard mitigation, new report says. IIPLR Update (June):1.

Insurance Service Office. 1994. Impact of catastrophes on property insurance. Executive summary. http://www.iso.com/docs/stud006.html.

Insurance Service Office. 1998. Building code enforcement report. http://www.iso.com:80/catalog/docs/016.html.

Interagency Ecosystem Management Task Force. 1995. The Ecosystem Approach: Healthy Ecosystems and Sustainable Economics. Washington, D.C.: The Task Force.

Interagency Floodplain Management Review Committee. 1994. Sharing the Challenge: Floodplain Management into the 21st Century. Washington, D.C.: U.S. Government Printing Office.

Interagency Hazard Mitigation Team. 1992. Interagency Hazard Mitigation Team Report in Response to the August 24, 1992 Disaster Declaration for the State of Florida. FEMA-955-Dr-FL. Atlanta, Ga.: Federal Emergency Management Agency.

Interagency Task Force on Floodplain Management. 1986. Unified Federal Program for Floodplain Management. Washington, D.C.: U.S. Government Printing Office.

Jacobs, Michael. 1991. The Green Economy: Environment, Sustainable Development and the Politics of the Future. London: Pluto Press.

Jaffe, Martin, JoAnn Butler, and Charles Thurow. 1981. Reducing Earthquake Risks: A Planner's Guide. PAS Report 364. Chicago: American Planning Association.

Jarrell, Jerry D., Paul J. Hebert, and Max Mayfield. 1992. Hurricane Experience Levels of Coastal County Populations from Texas to Maine. Washington, D.C.: National Technical Information Service.

Jarvinen, Brian R., and Miles B. Lawrence. 1985. An evaluation of the SLOSH storm-surge model. Bulletin American Meteorological Society 66(11):1408–1411.

John, DeWitt. 1994. Civic Environmentalism: Alternatives to Regulation in States and Communities. Washington, D.C.: Congressional Quarterly Press.

Johnson, F. Reed, Ann Fisher, V. Kerry Smith, and William H. Desvousges. 1988. Informed choice or regulated risk? Lessons from a study in radon risk communication. Environment 30(4):12–15, 30–35.

Kaiser, Edward J., and Matthew Goebel. 1996. Mitigation Following the 1993 Floods in Missouri. Chapel Hill: Center for Urban and Regional Studies, University of North Carolina.

Kaiser, Edward J., David R. Godschalk, and F. Stuart Chapin, Jr. 1995. Urban Land Use Planning. Urbana: University of Illinois Press.

Kaplan, John, and Mark DeMaria. 1995. A Simple Empirical Model for Predicting the Decay of Tropical Cyclone Winds after Landfall. Miami: National Oceanic and Atmospheric Administration/Atlantic Oceanographic and Meteorological Laboratories, Hurricane Research Division.

Kartez, Jack, and Charles Faupel. 1995. Factors promoting comprehensive local government hazards management. In From the Mountains to the Sea—Developing Local Capabilities. Proceedings of the Nineteenth Annual Conference of the Association of State Floodplain Managers. Natural Hazards Research and Applications Information Center Special Publication 31. Portland, Maine. Boulder: Institute of Behavioral Science, University of Colorado.

Keefer, David K. 1984. Landslides caused by earthquakes. Geological Society of America Bulletin 95:406–421.

Kellert, Stephen R. 1996. The Value of Life: Biological Diversity and Human Society. Washington, D.C.: Island Press.

Kelley, Joseph T., Alice R. Kelley, Orrin H. Pilkey, Sr., and Albert A. Clark. 1984. Living with the Louisiana Shore. Durham, N.C.: Duke University Press.

Kidd, Charles V. 1992. The evolution of sustainability. Journal of Agricultural and Environmental Ethics 5(1):1–26.

King, Mona. 1991. The economic benefits of hurricane and storm damage reduction. In Coastal Zone '91, Proceedings of the Seventh Symposium on Coastal and Ocean Management, Orville T. Magoon et al., eds. New York: American Society of Civil Engineers.

Krutilla, John V. 1966. An economic approach to coping with flood damage. Water Resources Research 2 (Second Quarter):183–190.

Kunruether, Howard. 1974. Economic analysis of natural hazards: An ordered choice approach. In Natural Hazards: Local, National, Global, Gilbert F. White, ed. New York: Oxford University Press.

Kunreuther, Howard. 1994. The role of insurance and mitigation in reducing disaster losses. In Natural Disasters: Local and Global Perspectives. Proceedings of the 1993 Annual Forum of the National Committee on Property Loss Reduction. Boston: Insurance Institute for Property Loss Reduction.

Kunreuther, Howard, Ralph Ginsberg, Louis Miller, Philip Sagi, Paul Slovic, Bradley Borkan, and Norman Katz. 1978. Disaster Insurance Protection: Public Policy Lessons. New York: John Wiley and Sons.

Kunster, James Howard. 1993. The Geography of Nowhere. New York: Touchstone.

Kusler, Jon A. 1980. Regulating Sensitive Lands. Cambridge, Mass.: Ballinger Publishing.

Kusler, Jon A. 1982. Innovation in Local Floodplain Management: A Summary of Community Experience. Natural Hazards Research and Applications Information Center Special Publication #4. Boulder: Institute of Behavioral Science, University of Colorado.

Kusler, Jon A. 1985. Regional programs meet national policy, roles along the rivers. Environment 27(7):18.

Kusler, Jon A., and Larry Larson. 1993. Beyond the ark: A new approach to U.S. floodplain management. Environment 35(June):7–34.

L. R. Johnston and Associates. 1991. Status Report on the Nation's Floodplain Management Activities. Draft. Washington, D.C.: Federal Emergency Management Agency.

L. R. Johnston and Associates. 1992. Floodplain Management in the United

States: An Assessment Report, volume 2: Full Report. Washington, D.C.: U.S. Government Printing Office.

Lambright, W. Henry. 1984. Role of States in Earthquake and Natural Hazard Innovation at the Local Level: A Decision-Making Study. Washington, D.C.: National Technical Information Service.

Land Trust Alliance. 1991. Land trusts have helped to protect 2.7 million acres. Land Trust Exchange 10(2):15–16.

Land Trust Alliance. 1995. Directory of Land Trusts Involved in Floodplain or Coastal Protection. Washington, D.C.: The Alliance.

Landis, John D. 1995. Imagining land use futures: applying the California Urban Futures Model. Journal of the America Planning Association 61(Autumn):438–456.

Langdon, Phillip. 1994. A Better Place to Live. New York: Harper Press.

Laska, Shirley Bradway. 1991. Floodproof Retrofitting: Homeowner Self-Protective Behavior. Boulder: Institute of Behavioral Science, University of Colorado.

Lecomte, Eugene L. 1992. Professional organizations in hazards mitigation and management: How effective? Paper presented at the 17th Annual Hazards Research and Applications Workshop, Boulder, Colo., July 12–15.

Lecomte, Eugene L. 1995. Remarks on mitigation. National Mitigation Conference, Arlington, Va., December 7.

Lee County Division of Public Safety. 1993. Emergency Public Shelter Impact Fee Proposal. Ft. Myers, Fla.: The Division.

Lees, Annette. 1995. Innovative partners: The value of nongovernmental organizations in establishing and managing protected areas. In Expanding Partnerships in Conservation, Jeffrey A. McNeely, ed. Washington, D.C.: Island Press.

Leinberger, Christopher B. 1996. Metropolitan development trends of the late 1990's: Social and environmental implications. In Land Use in America, Henry L Diamond and Patrice F. Noonan, eds. Washington, D.C.: Island Press.

Leyendecker, E. V., D. M. Perkins, S. T. Algermissen, P. C. Thenhaus, and S. L. Hanson. 1995. USGS Spectral Response Maps and Their Relationship with Seismic Design Forces in Building Codes. Open-File Report 95–596. Reston, Va.: U.S. Geological Survey.

Lichtenberg, Erik. 1994. Sharing the challenge: An economist's view. Water Resources Update 97(Autumn):39–43.

Linn, Steven. 1988. Flood Management Study of the C-18 Basin. West Palm Beach, Fla..: South Florida Water Management District.

Litan, Robert, Frederick Krimgold, Karen Clark, and Jayant Khadilkar. 1992. Physical Damage and Human Loss: The Economic Impact of Earthquake Mitigation Measures. Boston: The National Committee on Property Insurance.

Little, Charles E. 1995. Greenways for America. Baltimore, Md.: Johns Hopkins University Press.

Local Government Commission. 1995. Participation Tools for Better Land Use Planning. Sacramento, Calif.: The Commission.

Logan, John R., and Harvey Molotch. 1987. Urban Fortunes: The Political Economy of Place. Berkeley: University of California Press.

Los Angeles, City of. 1994. Recovery and Reconstruction Plan. Los Angeles: Emergency Operations Organization, City of Los Angeles.

Los Angeles County. 1990. Safety Element, Los Angeles County General Plan. Los Angeles: County Department of Regional Planning.

Louisiana Coastal Wetlands Conservation and Restoration Task Force. 1993. Louisiana Coastal Wetlands Restoration Plan: Main Report and Environmental Impact Statement. Baton Rouge, La.: The Task Force.

Lovelace, Thomas, and Thomas S. Cantine. 1977. Management—institutional aspects of water quality management planning. In Handbook of Water Quality Management, J. L. Pavoni, ed. New York: Van Nostrand-Reinhold.

Mader, George C., and Martha B. Tyler. 1993. Land use and planning. In Improving Earthquake Mitigation, Report to Congress. Washington, D.C.: Federal Emergency Management Agency.

Mantell, Michael A., Stephen F. Harper, and Luther Propst. 1990. Creating Successful Communities: A Guidebook to Growth Management Strategies. Washington, D.C.: Island Press.

Maser, Christopher. 1995. Resolving Environmental Conflict: Towards Sustainable Community Development. Delray Beach, Fla.: St. Lucie Press.

Mawlawi, Farouk. 1993. New conflicts, new challenges: The evolving role for NGO actors. Journal of International Affairs 46(2):391–413.

May, Peter J. 1989. Anticipating Earthquakes: Risk Reduction Policies and Practices in the Puget Sound and Portland Areas. Seattle, Wash.: Institute for Public Policy and Management, University of Washington.

May, Peter J. 1991. Addressing public risks: Federal earthquake policy design. Journal of Policy Analysis and Management 10(Spring):263–285.

May, Peter J. 1994. Analyzing mandate design: State mandates governing hazard-prone areas. Publius, The Journal of Federalism 24(Spring 1994):1–16.

May, Peter J. 1996. Addressing natural hazards: Challenges and lessons for public policy. Australian Journal of Emergency Management 11(Summer):30–37.

May, Peter J., and Thomas A. Birkland. 1994. Earthquake risk reduction: An examination of local regulatory efforts. Environmental Management 18(November/December):923–937.

May, Peter J., and Patricia T. Bolton. 1986. Reassessing earthquake hazards reduction measures. Journal of the American Planning Association 52(Autumn):443–454.

May, Peter J., and Raymond J. Burby. 1996. Coercive versus cooperative policies: Comparing intergovernmental mandate performance. Journal of Policy Analysis and Management 15(2):171–201.

May, Peter J., and Nancy Stark. 1992. Design professions and earthquake policy. Earthquake Spectra 8(February):115–132.

May, Peter J., and Walter Williams. 1986. Disaster Policy Implementation: Managing Programs Under Shared Governance. New York: Plenum Publishing.

May, Peter J., Raymond J. Burby, Neil Ericksen, John Handmer, Jennifer Dixon, Sarah Michaels, and D. Ingle Smith. 1996. Environmental Management and Governance: Intergovernmental Approaches to Hazards and Sustainability. London and New York: Routledge Press.

McCreary, Scott T., and John R. Clark. 1983. Community flood hazard management for the coastal barriers of Apalachicola Bay, Florida. In Preventing Coastal Flood Disasters: The Role of the States and Federal Response, Proceedings of a National Symposium, Jacquelyn Monday, ed. Natural Hazards Research and Applications Information Center Special Publication #7. Boulder: Institute of Behavioral Science, University of Colorado.

McHarg, Ian. 1969. Design with Nature. Garden City, N.J.: Anchor Books.

McLoughlin, David. 1985. A framework for integrated emergency management. Public Administration Review 45(Special Issue, January):165–172.

Meadows, Donnella H., Dennis L. Meadows, and Jorgen Randers. 1992. Beyond the Limits. New York: Chelsea Green.

Meltsner, Arnold J. 1978. Public support for seismic safety: Where is it in California? Mass Emergencies (3):167–184.

Metropolitan Dade County, Florida. 1966. Application Requesting Amendment to the Coastal Management Element. Metro-Dade County Comprehensive Development Master Plan. Miami, Fla.: Metro-Dade County Department of Planning, Development, and Regulation.

Miami Herald. December 20, 1992, p. 3.

Michaels, Sarah M. 1992. New perspectives on diffusion of earthquake knowledge. Earthquake Spectra 8(1):159–175.

Miller, Donald, and Frank Westerlund. 1990. Specialized land use curricula in urban planning graduate programs. Journal of Planning Education and Research 9(3):203–206.

Milliman, Jerome W. 1983. An agenda for economic research on flood hazard mitigation. In A Plan for Research on Floods and Their Mitigation in the United States, Stanley Chagnon, ed. Champaign, Ill.: Illinois State Water Survey.

Minear, Larry. 1987. The other missions of NGOs: Education and advocacy. World Development 15(Supplement):201–211.

Minnesota Environmental Quality Board. 1995. Common Ground: Achieving Sustainable Communities in Minnesota. Report of the Sustainable Economic

Development and Environment Task Force to the Governor, the Minnesota Legislature, and the Minnesota Environmental Quality Board. September. St. Paul, Minn.: The Board.

Morris, Wendy, and J. R. Barber. 1982. Fire Hazard Mapping. Victoria, Australia: County Fire Authority.

Mullin, John R. 1992. The reconstruction of Lisbon following the earthquake of 1755: A study in despotic planning. Planning Perspectives 7:157–179.

Munasinghe, Mohan, and Caroline Clarke, eds. 1995. Disaster Prevention for Sustainable Development: Economic and Policy Issues. Washington, D.C.: The International Bank for Reconstruction and Development/The World Bank.

Murphy, Francis C. 1958. Regulating Flood Plain Development. Department of Geography Research Paper No. 56. Chicago: Department of Geography, University of Chicago.

Murray, Will. 1995. Lessons from 35 years of private preserve management in the USA: The preserve system of the Nature Conservancy. In Expanding Partnerships in Conservation, Jeffrey A. McNeely, ed. Washington, D.C.: Island Press.

Mushkatel, Alvin H., and Joanne M. Nigg. 1987a. Effect of objective risk on key actor support for seismic mitigation policy. Environmental Management 11(1):77–86.

Mushkatel, Alvin H., and Joanne M. Nigg. 1987b. Opinion congruence and the formulation of seismic safety policies. Policy Studies Review 6:645–656.

Nabhan, Gary Paul, and Stephen Trimble. 1994. The Geography of Childhood. Boston: Beacon Press County.

Nags Head, N.C., City of. 1990. Land Use Plan for Nags Head, North Carolina. Nags Head, N.C.: Planning Department, City of Nags Head.

National Academy of Public Administration. 1993. Coping with Catastrophe: Building on Emergency Management Systems to Meet People's Needs in Natural and Manmade Disasters. February. Washington, D.C.: The Academy.

National Academy of Sciences. 1977. Methodology for Calculating Wave Action Effects Associated with Storm Surges. Washington, D.C.: The Academy.

National Commission on the Environment. 1993. Choosing A Sustainable Future. Washington, D.C.: Island Press.

National Institute of Building Sciences. 1994. Development of a Standardized Earthquake Loss Estimation Methodology, volumes I and II. Washington, D.C.: The Institute.

National Park Service. 1991. A Casebook in Managing Rivers for Multiple Uses. Washington, D.C.: NPS.

National Performance Review. 1993. From Red Tape to Results: Creating a

Government that Works Better and Costs Less. Washington, D.C.: U.S. Government Printing Office.

National Research Council. 1990. Managing Coastal Erosion. Washington, D.C.: National Academy Press.

National Research Council, Commission on Geosciences, Environment and Resources. 1991. A Safer Future: Reducing the Impacts of Natural Disasters. Washington, D.C.: National Academy Press.

National Research Council. 1993. Wind and the Built Environment: U.S. Needs in Wind Engineering and Hazard Mitigation. Washington, D.C.: National Academy Press.

National Science and Technology Council. 1996. Natural Disaster Reduction: A Plan for the Nation. February. Washington, D.C.: Subcommittee on Natural Disaster Reduction, National Science and Technology Council. February.

Neal, David M. 1990. Volunteer Organization Responses to the Earthquake. In The Loma Prieta Earthquake: Studies of Short-Term Impacts, Robert Bolin, ed. Boulder: Institute of Behavioral Science, University of Colorado.

Nelessen, Anton C. 1994. Visions for a New American Dream. Chicago: American Planning Association, Planners Press.

Nelson, Arthur C., and James B. Duncan. 1995. Growth Management Principles and Practices. Chicago: American Planning Association, Planners Press.

Neumann, Charles J. 1987. The National Hurricane Risk Analysis Program. NOAA Technical Memorandum NWS NHC 38. Coral Gables, Fla.: National Hurricane Center.

Newman, Peter. 1996. Greening the city. Alternatives Journal 22(2):10–16.

Norris, Ruth, and Laura Camposbasso. 1995. Protected areas and the private sector: Building NGO relationships. In Expanding Partnerships in Conservation, Jeffrey A. McNeely, ed. Washington, D.C.: Island Press.

North Carolina, State of. 1996. North Carolina Administrative Code 15A Subchapter 7B - Land Use Planning Guidelines, Section .0212 (a) (5) Storm Hazard Mitigation, Post-Disaster Recovery and Evacuation Plans. Raleigh, N.C.: State of North Carolina.

Noss, Reed F., Edward T. Lafoe III, and Michael Scott. 1995. Endangered Ecosystems of the United States: A Preliminary Assessment of Loss and Degradation. Washington, D.C.: U.S. Fish and Wildlife Service.

Olshansky, Robert B. 1990. Landslide Hazard in the United States: Case Studies in Planning and Policy Development. New York: Garland Publishing.

Olshansky, Robert B. 1994. Seismic Hazard in the Central United States: The Role of the States. Professional Paper 1538-G. Reston, Va.: U.S. Geological Survey.

Olshansky, Robert B. 1996. The California Environmental Quality Act and local planning. Journal of the American Planning Association 62(3):313–330.

Olshansky, Robert B., Howard Foster, and Stuart Cook. 1991. Microzonation: A planner's perspective. In Proceedings of the Fourth International Conference on Seismic Zonation, volume III. Oakland, Calif.: Earthquake Engineering Research Institute.

Orians, Carlyn E., and Patricia A. Bolton. 1992. Earthquake Mitigation Programs in California, Utah and Washington. BHARC-800/92/041. Seattle, Wash.: Battelle Human Affairs Research Center.

Palm, Risa. 1981. Real Estate Agents and Special Studies Zone Disclosure: The Response of California Home Buyers to Earthquake Hazards Information. Boulder: Institute of Behavioral Science, University of Colorado.

Panel on Earthquake Loss Estimation Methodology, Committee on Earthquake Engineering, Commission on Engineering and Technical Systems, National Research Council. 1989. Estimating Losses from Future Earthquake—Panel Report. Washington, D.C.: National Academy Press.

Patel, Jigar. 1991. Estimating Infrastructure Elements Using Geographic Information Systems. Ithaca, N.Y.: Institute for Social and Economic Research Program in Urban and Regional Studies, Cornell University.

Paterson, Robert G. 1996. Land Trusts: Community Catalysts for Floodplain Management. Austin, Tex.: College of Architecture, University of Texas.

Patton, Ann. 1994. From Rooftop to River: Tulsa's Approach to Floodplain and Stormwater Management. Tulsa, Okla.: Public Works Department, City of Tulsa.

Perkins, Jeanne B. 1992. Estimates of Uninhabitable Dwelling Units in Future Earthquake Affecting the San Francisco Bay Region. Oakland, Calif.: Association of Bay Area Governments.

Petak, William J., and Arthur A. Atkisson. 1982. Natural Hazard Risk Assessment and Public Policy: Anticipating the Unexpected. New York: Springer-Verlag.

Phillippi, Nancy. 1994-95. Plugging the Gaps in Flood Control Policy, Issues in Science and Technology (Winter):71–78.

Pilkey, Orrin. 1980. From Currituck to Calabash: Living with North Carolina's Barrier Islands. Durham, N.C.: Duke University Press.

Plafker G., and J. P. Galloway, eds. 1989. Lessons Learned from the Loma Prieta, California, Earthquake of October 17, 1989. U.S. Geological Survey Circular 1045. Washington, D.C.: U.S. Government Printing Office.

Platt, Rutherford H. 1982. The Jackson flood of 1979: A public policy disaster. Journal of the American Planning Association 48(Spring):219–231.

Platt, Rutherford H. 1986. Metropolitan flood loss reduction through special regional districts. Journal of the American Planning Association 52(Autumn):467–479.

Platt, Rutherford H., ed. 1987. Regional Management of Metropolitan Floodplains. Experience in the United States and Abroad. Boulder: Institute of Behavioral Science, University of Colorado.

Platt, Rutherford H. 1992. An eroding base. Environmental Forum 9(5):10–15.

Platt, Rutherford H. 1995. Report on reports: Sharing the challenge: Floodplain management into the 21st century. Environment 37(1):25–29.

Platt, Rutherford H. 1996. Land Use and Society: Geography, Law, and Public Policy. Washington, D.C.: Island Press.

Power, Thomas Michael. 1996. The wealth of nature. Issues in Science and Technology 12(Spring):48–54.

President's Council on Sustainable Development. 1996. Sustainable America: A New Consensus for Prosperity, Opportunity, and a Healthy Environment. Washington, D.C.: U.S. Government Printing Office.

Prestby, John E., Abraham Wandersman, Paul Florin, Richard Rich, and David Chavis. 1990. Benefits, costs, incentive management and participation in voluntary organizations: A means to understanding and promoting empowerment. American Journal of Community Psychology 18(1):117–150.

Quarantelli, Enrico L., et al. 1983. Emergent Citizen Groups in Disaster Preparedness and Recovery Activities. Miscellaneous Report 33. Columbus, Ohio: Disaster Research Center, Ohio State University.

Radke, John. 1995. Modeling urban/wildland interface fire hazards within a geographic information system. Geographic Information Sciences 1(1):9–22.

Rasmussen, Steen Eiler. [1934] 1967. London: The Unique City. Cambridge, Mass.: MIT Press.

Real, Charles R., and Michael S. Reichle. 1995. An accelerated program for seismic zonation following the January 17, 1994 Northridge earthquake. In Proceedings of the Fifth International Conference on Seismic Zonation. Nantes, France: Ouest Editions.

Rees, William E. 1992. Ecological footprints and appropriated carrying capacity: What urban economics leaves out. Environment and Urbanization 4(2):121–30.

Rees, William E., and Mark Roseland. 1991. Sustainable Communities: Planning for the 21st Century. Plan Canada 31:15–26.

Roelofs, Joan. 1995. The third sector as a protective layer for capitalism. Monthly Review 47(4):16–26.

Ronfeldt, David, and Cathryn Thorup. 1995. North America in the Era of Citizen Networks: State, Society and Security. Santa Monica, Calif.: Rand Corp.

Roseland, Mark. 1992. Toward Sustainable Communities. Ottawa, Canada: National Roundtable on Environment and Economy.

Rossi, Peter, James D. Wright, and Eleanor Weber-Burdin. 1982. Natural Hazards and Public Choice: The State and Local Politics of Hazard Mitigation. New York: Academic Press.

Rubin, Claire B., and Roy Popkin. 1990. Disaster Recovery After Hurricane Hugo in South Carolina. Working Paper #69. Boulder: Institute of Behavioral Science, University of Colorado.

Rubin, Claire B., Martin D. Saperstein, and Daniel G. Barbee. 1985. Community Recovery from a Major Natural Disaster. Boulder: Institute of Behavioral Science, University of Colorado.

Ruch, Carl. 1983. Hurricane Vulnerability Analysis for Brazoria, Galveston, Harris, and Chambers Counties. College Station, Tex.: College of Architecture and Environmental Design, Texas A&M University.

Russell, James E. 1994. Post-earthquake reconstruction regulation by local government. Earthquake Spectra 10(1):209–223.

Saffir, Herbert S. 1992. An evaluation of present-day hurricane-resistant building codes. Journal of Coastal Research 8(2):492–495.

Salamon, Lester M. 1989. The changing partnership between the voluntary sector and the welfare state. In The Future of the Nonprofit Sector: Challenges, Changes and Policy Considerations, V. A. Hodgkinson and R. W. Lyman, eds. San Francisco: Jossey-Bass.

Salamon, Lester M. 1994. The rise of the nonprofit sector. Foreign Affairs 73:109–25.

Sale, Kirkpatrick. 1991. Dwellers in the Land: The Bioregional Vision. Philadelphia, Pa.: New Society Publishers.

Sampson, David. 1996. Hudson River Greenway: A regional success story. In Land Use in America, Henry L. Diamond and Patrick F. Noonan, eds. Washington, D.C.: Island Press.

Sandman, Peter M. 1985. Getting to maybe: Some communications aspects of siting hazardous waste facilities. Seton Hall Legislative Journal 9:442–465.

San Francisco Planning Department. 1996a. Community Safety Element. Draft for Citizen Review. An Element of the Master Plan of the City and County of San Francisco. San Francisco: The Department.

San Francisco Planning Department. 1996b. Community Safety Element Summary Background Report. San Francisco: The Department.

Sato, Ryo, and Masanobu Shinozuka. 1991. GIS-based interactive and graphic computer system to evaluate seismic risks on water delivery networks. In Lifeline Earthquake Engineering: Proceedings of 3rd U.S. Conference August 22–23, 1991, Michael A. Cassaro ed. New York: American Society of Civil Engineers.

Schneider, Saundra K. 1995. Flirting with Disaster: Public Management in Crisis Situations. Armonk, N.Y.: M. E. Sharpe.

Schulz, Paula A. 1993. Education, awareness, and information transfer issues. In Improving Earthquake Mitigation: Report to Congress. Washington, D.C.: Federal Emergency Management Agency.

Schuster, Robert L., and Robert W. Fleming. 1986. Economic losses and fatalities due to landslides. Bulletin of the Association of Engineering Geologists 23(1):11–28.

Schwerdt, Richard W., Francis P. Ho, and Roger R. Watkins. 1979. Meteorological Criteria for Standard Project Hurricane and Probable Maximum Hurricane Windfields, Gulf and East Coasts of the United States. NOAA Technical Report NWS 23. Washington, D.C.: National Weather Service, U.S. Department of Commerce.

Science Applications International Corporation (SAIC). 1989. A Program for Assessing Hurricane Risk (HURISK), Revision 1. McLean, Va.: SAIC.

Scott, Jacquelyn T. 1995. Some thoughts on theory development in the voluntary and nonprofit sector. Nonprofit and Voluntary Sector Quarterly 24(1)31–41.

Seligson, H. A., et al. 1990. Comparison of lifeline response and recovery performance in the 1971 San Fernando and 1987 Whittier Narrows earthquakes. In Proceedings of Fourth U.S. National Conference on Earthquake Engineering, May 20-24, volume 1. El Cerrito, Calif.: Earthquake Engineering Research Institute.

Shah, Haresh C. 1994. Scientific profiles of 'the big one.' Natural Hazards Observer 20, 2(November):1–3.

Sheaffer, John R., et al. 1976. Flood Hazard Mitigation Through Safe Land Use Practices. Chicago: Kiefer and Associates.

Simpson, David M. 1993. Risky business: natural hazards and the field of planning. Journal of Planning Education and Research 12(3):252–254.

Simpson, Robert H., and Herbert Riehl. 1980. The Hurricane and Its Impact. Baton Rouge: Louisiana State University Press.

Singh, J. V., and C. J. Lumsden. 1990. Theory and research in organizational ecology. Annual Review of Sociology 16:161–195.

Skolnick, Richard. 1993. Rebuilding trust: Nonprofits act to boost reputations. Public Relations Journal 49:29–35.

Slovic, Paul, Baruch Fischhoff, and Sarah Lichtenstein. 1979. Rating the risks. Environment 21(3):14–20, 36–39.

Slovic, Paul, Baruch Fischoff, and Sarah Lichtenstein. 1987. Behavioral decision theory perspectives on protective behavior. In Taking Care: Understanding and Encouraging Self-Protective Behavior, Neil D. Weinstein, ed. Cambridge, England: Cambridge University Press.

Smith, Daniel S., and Paul Cawood Hellmund, eds. 1993. Ecology of Greenways. Minneapolis, Minn.: University of Minnesota Press.

Smith, Di. 1994. Flood damage estimation—A review of urban stage-damage curves and loss functions. Water SA 20(3):231–238.

Smith, John A. 1989. The evolving role of foundations. In The Future of the Nonprofit Sector: Challenges, Changes and Policy Considerations, V. A. Hodgkinson and R. W. Lyman, eds. San Francisco: Jossey-Bass.

Smith, Keith L. 1995. Main Street at 15. Historic Preservation Forum 9(3):59–63.

Smith, Richard A. 1995. Hurricane Loss Damage Curves. Tallahassee, Fla.: Florida Planning Laboratory, Department of Urban and Regional Planning, Florida State University.

Smith, Richard A., and Robert E. Deyle. 1994. Development of a Risk-Based Mechanism for Funding Local Government Coastal Storm Hazard Management Services. Tallahassee, Fla.: Department of Urban and Regional Planning, Florida State University.

Smith, Richard A., and Robert E. Deyle, with Alexander Gallagher, Chris Killingsworth, and Greg Williamson. 1996. The Use and Impact of Local Hazard Mitigation and Post-Storm Redevelopment Plans in the Aftermath of Hurricanes: Lessons Learned from Opal. Tallahassee, Fla.: The Florida Planning Laboratory, Department of Urban and Regional Planning, Florida State University.

South Florida Water Management District. 1984. Water Management Planning for the Western C-51 Basin, Palm Beach County, Florida. West Palm Beach, Fla.: The District.

Southern Building Code Congress International. 1992. Coastal Building Department Survey. Chicago: Natural Disaster Loss Reduction Committee, National Committee on Property Insurance.

Steinbrugge, Karl V., and S. T. Algermissen. 1990. Earthquake Losses to Single-Family Dwellings: California Experience. U.S. Geological Survey Bulletin 1939-A. Washington, D.C.: U.S. Government Printing Office.

Steinbrugge, Karl V., John H. Bennett, Henry J. Lagorio, James F. Davis, Glenn Borchardt, and Tousson R. Toppozada. 1987. Earthquake Planning Scenario for a Magnitude 7.5 Earthquake on the Hayward Fault in the San Francisco Bay Area. Special Publication 78. Sacramento, Calif.: California Division of Mines and Geology.

Sustainable Economic Development and Environmental Protection Task Force. 1995. Common Ground: Achieving Sustainable Communities in Minnesota. September. St. Paul, Minn.: The Task Force.

Sustainable Seattle. 1995. Indicators of Sustainable Community. Seattle, Wash.: Sustainable Seattle.

Swink, Rodney. 1995. Partnerships: The organizational model for challenging times. Historic Preservation Forum 9(2):2–8.

Talen, Emily. 1996. Do plans get implemented? A review of evaluation in planning. Journal of Planning Literature 10(3):248–259.

Tampa Bay Regional Planning Council. 1993. Tampa Bay Region Hurricane Loss and Contingency Planning Study. Tampa, Fla.: The Council.

Tennessee Emergency Management Agency. No date. Hazard Mitigation Planning Guidance for Local Governments. Nashville, Tenn.: The Office.

Tennessee Local Planning Assistance Office. No date. Local Hazard Mitigation Planning: Format, Forms, and Example. Chattanooga, Tenn.: The Agency.

Thieler, Edward R., and David M. Bush. 1991. Hurricanes Gilbert and Hugo send powerful messages for coastal development. Journal of Geological Education 39:291–299.

Thomas, G., and M. M. Witts. 1971. The San Francisco Earthquake. New York: Stein and Day.

Tierney, Kathleen J. 1981. Community and organizational awareness of and preparedness for acute chemical emergencies. Journal of Hazardous Materials 4:331–342.

Tobin, L. Thomas. 1991. California Urban Hazard Mapping Program: A bold experiment in earth science and public policy. In Proceedings, Fourth International Conference on Seismic Zonation, August 25–29, volume 3. Oakland, Calif.: Earthquake Engineering Research Institute.

Tobin, Graham A., and Jane C. Ollenburger. 1992. National Hazards and the Elderly. Quick Response Research Report #53. Boulder: Institute of Behavioral Science, University of Colorado.

Trust for Public Lands. 1995. Doing Deals: A Guide to Buying Land for Conservation. Washington, D.C.: The Land Trust Alliance.

United Nations. 1992. Agenda 21. New York: United Nations.

U.S. Advisory Commission on Intergovernmental Relations. 1993. Federal Regulation of State and Local Governments: The Mixed Record of the 1980s. Washington, D.C.: The Commission.

U.S. Army Corps of Engineers, Mobile District. 1988. Tri-State Hurricane Property Loss Study: Technical Data Report. Mobile, Ala.: The District.

U.S. Army Corps of Engineers, Mobile District. 1990. Tri-State Hurricane Property Loss Study: Executive Summary and Technical Data Report. Mobile, Ala.: The District.

U.S. Army Corps of Engineers, Mobile District. 1994. Panama City Beaches, Florida, Beach Erosion Control and Storm Damage Reduction Project, Appendix C: Economic Investigations. Mobile, Ala.: The District.

U.S. Bureau of the Census. 1991. 1990 Census of Population and Housing: Summary Tape File 1B Extract on CD-ROM, Technical Documentation. Washington, D.C.: Bureau of the Census, U.S. Department of Commerce.

U.S. Congress, General Accounting Office. 1990a. Federal-State-Local Relations, Trends of the Past Decade and Emerging Issues. Report HRD-90-34. Washington, D.C.: GAO.

U.S. Congress, General Accounting Office. 1990b. Flood Insurance Information on the Mandatory Purchase Requirement. Report No. RCED-90-141FS, August 22. Washington, D.C.: GAO.

U.S. Congress, General Accounting Office. 1992. Coastal Barriers: Development Occurring Despite Prohibitions Against Federal Assistance. Report to the Committee on Environment and Public Works, U.S. Senate. Washington, D.C.: GAO.

U.S. Congress, General Accounting Office. 1993. Disaster Management: Improving the Nation's Response to Catastrophic Disasters. Washington, D.C.: GAO.

U.S. Congress, General Accounting Office. 1995. EPA and the States: Environmental Challenges Require a Better Working Relationship. Report RCED-95-64. Washington, D.C.: GAO.

U.S. Environmental Protection Agency. 1992. Private Landowner's Wetlands Assistance Guide: Voluntary Options for Wetlands Stewardship in Maryland. Washington, D.C.: EPA.

U.S. Environmental Protection Agency, Office of Water. 1993. Using Nonprofit Organizations to Advance Estuary Goals. Washington, D.C.: EPA.

U.S. Environmental Protection Agency, Office of Water. 1994. The Watershed Protection Approach. Washington, D.C.: EPA.

U.S. Geological Survey, Working Group on California Earthquake Probabilities. 1990. Probabilities of Large Earthquakes in the San Francisco Bay Region, California. U.S. Geological Survey Circular 1053. Washington, D.C.: U.S. Government Printing Office.

U.S. Geological Survey. 1994. The Next Big Earthquake in the Bay Area May Come Sooner Than You Think. (Public Information Pamphlet.) Menlo Park, Calif.: USGS.

U.S. Geological Survey, Working Group on California Earthquake Probabilities. 1995. Seismic hazards in southern California: probable earthquakes, 1994–2024. Bulletin of the Seismological Society of America 85(2):379–439.

U.S. House of Representatives. 1994. Report of the Bipartisan Task Force on Disasters. December 14. Washington, D.C.: Congress of the United States, House of Representatives.

U.S. National Committee for the Decade for Natural Disaster Reduction. 1991. A Safer Future: Reducing the Impacts of Natural Disasters. Washington, D.C.: National Academy Press.

U.S. Senate Task Force on Funding Disaster Relief. 1995. Federal Disaster Assistance. Senate Document 104-4. Washington, D.C.: U.S. Government Printing Office.

Uphoff, Norman. 1986. Local Institutional Development: An Analytic Sourcebook With Cases. West Hartford, Conn.: Kumarian Press.

Uphoff, Norman. 1993. Grassroots organizations and NGOs in rural development: Opportunities with diminishing states and expanding markets. World Development 21(4):607–622.

Upper Arkansas Valley Wildfire Council. 1993a. Annual Report. Salida, Colo.: Chaffee Press.

Upper Arkansas Valley Wildfire Council. 1993b. Home Wildfire Protection Guide. Salida, Colo.: Chaffee Press.

Van der Ryn, Sim, and Peter Calthorpe. 1991. Sustainable Communities. San Francisco: Sierra Club.

Van Til, Jon. 1988. Mapping the Third Sector: Volunteerism in a Changing Social Economy. Washington, D.C.: The Foundation Center.

Varnes, David J. 1984. Landslide Hazard Zonation: A Review of Principles and Practice. Paris: UNESCO.

Viessman, William, and Charles Welty. 1985. Water Management Technology and Institutions. New York: Harper and Row.

Vonier, Thomas. 1993. Blown away: The aftermath of Hurricane Andrew. Progressive Architecture 74(6):124–128.

Wackernagel, Mathis, and William Rees. 1995. Our Ecological Footprint: Reducing Impact on the Earth. Gabriola Island, British Columbia, Canada: New Society Publishers.

Wahl, Richard W. 1994. The Mississippi Flood. Environment 36(5):2–12.

Waldo, Dwight. 1973. Epilogue: Public service professional associations in context of socio-political transition. In Public Service Professional Associations and Public Interest Groups, Don L. Bowen, ed. Philadelphia, Pa.: American Academy of Political Science.

Wall, S. 1995. National Trust Summary of Activities Associated with Hurricanes Hugo and Andrew. (Database printout.) Charleston, S.C.: Southern Regional Office of the National Trust for Historic Preservation.

Wapner, Paul. 1995. Politics beyond the state: Environmental and world civic politics. World Politics 47:311–347.

Watson, C. C. 1995. The arbiter of storms: A high resolution, GIS based system for integrated storm hazard modeling. National Weather Digest 20(2):2–9.

Waugh, William L. 1990. Emergency management: State and local government capacity. In Cities and Disaster, R.T. Sylves and W.L. Waugh, eds. Springfield, Ill.: Charles C Thomas Publishing.

Weinstein, Milton C., and Robert J. Quinn. 1983. Psychological considerations in valuing health risk reduction. Natural Resources Journal 23(July):659–673.

Wernly, D. R. 1994. Mitigation Efforts. Aware: National Weather Service/ Warning Coordination and Hazard Awareness Report (Spring Issue). Washington, D.C.: National Oceanic and Atmospheric Administration, U.S. Department of Commerce.

Wetmore, French. 1986. Flood protection planning assistance to small towns. Planning and Public Policy 12(3):1–4.

White, Gilbert. 1945. Human Adjustment to Floods. Research Paper 29. Chicago: Department of Geography, University of Chicago.

White, Gilbert F. 1969. Strategies of American Water Management. Ann Arbor: University of Michigan Press.

White, Gilbert F. 1996. Emerging issues in global environmental policy. Ambio 25(1) (February):58–60.

White, Gilbert F., and J. Eugene Haas. 1975. Assessment of Research on Natural Hazards. Cambridge, Mass.: MIT Press.

White, Gilbert F., Wesley C. Calef, James W. Hudson, Harold M. Mayer, John R. Sheaffer, and Donald J. Volk. 1958. Changes in Urban Occupance of Flood Plains in the United States. Chicago: Department of Geography, University of Chicago.

Williamson, David. 1994. The Nature Conservancy. Environmental Planning Quarterly 10(4):5–8.

Wilson, Catherine. 1992. Construction flaws revealed in Hurricane Andrew's wake. New Orleans Times Picayune, August 28:A–6.

Wilson, Raymond C., and David K. Keefer. 1985. Predicting areal limits of earthquake-induced landsliding. In Evaluating Earthquake Hazards in the Los Angeles Region, J.I. Ziony, ed. Washington, D.C.: U.S. Government Printing Office, U.S. Geological Survey Professional Paper 1360.

Withlacoochee Regional Planning Council. 1987. Withlacoochee Region Hurricane Loss Study: Final Report. Ocala, Fla.: The Council.

Wolensky, Robert P., and Kenneth C. Wolensky. 1990. Local government's problem with disaster management: A literature review and structural analysis. Policy Studies Review 9:703–725.

World Commission on Environment and Development. 1987. Our Common Future. Oxford, England: Oxford University Press.

Wright, James D., Peter H. Rossi, Sonia R. Wright, and Eleanor Weber-Burdin. 1979. After the Clean-Up: Long-Range Effects of Natural Disasters. Beverly Hills, Calif.: Sage Publications.

Wyner, Alan J., and Dean E. Mann. 1986. Preparing for California's Earthquakes: Local Government and Seismic Safety. Berkeley, Calif.: Institute for Governmental Studies, University of California.

Young, Dennis R. 1991. The structural imperatives of international advocacy associations. Human Relations 44(9):921–954.

Index

A

B

D

E

F

G

H

I

J

M

N

Q

R

S

T

Y

Z